Decentralized Control of Large-Scale Systems

Edward J. Davison · Amir G. Aghdam ·
Daniel E. Miller

Decentralized Control of Large-Scale Systems

Springer

Edward J. Davison
The Edward S. Rogers Sr. Department
of Electrical and Computer Engineering
The University of Toronto
Toronto, ON, Canada

Amir G. Aghdam
Department of Electrical and Computer Engineering
Concordia University
Montreal, QC, Canada

Daniel E. Miller
Department of Electrical
and Computer Engineering
University of Waterloo
Waterloo, ON, Canada

The solutions for problems and exercises will be found on the https://www.springer.com/in/book/9781441960139.

ISBN 978-1-4419-6013-9 ISBN 978-1-4419-6014-6 (eBook)
https://doi.org/10.1007/978-1-4419-6014-6

This Springer imprint is published by the registered company Springer Science+Business Media, LLC part of Springer Nature.
The registered company address is: 233 Spring Street, New York, NY 10013, U.S.A.

Ted Davison dedicates the book to his wife Zofia and his four children Laura, Daniel, Susan and Anne

Amir Aghdam dedicates the book to his wife Kamelia and his two children Tara and Tina

Daniel Miller dedicates the book to his wife Marilou

Preface

Whenever one sets out to achieve a new goal, one must have a starting point, and in this case, it will be assumed that the reader has a basic knowledge of classical control theory and a basic knowledge of the fundamentals of state-space theory. We will start off at the beginning of the topic of large-scale decentralized control systems and conclude with some of the latest results obtained to date. Have fun!

The Development of Decentralized Control

The topic of control theory arose during the development of the steam engine, and rapidly advanced during the 1940s–1950s period under the name of classical control, which deals mainly with "single-input single-output" systems. Kalman thence made a major contribution by introducing state-space methods, which allowed for the analysis of, and controller synthesis for, multi-input multi-output systems. As a consequence, control theory developed rapidly, but it was soon recognized that the theory was hard to apply to very complex systems, such as a large chemical plant or a large power system: in these cases, there are typically information flow constraints, often arising from the spatial separation of various sensors and actuators. Ad hoc short cuts were adopted to circumvent this problem, which sometimes worked, but often failed. This led to the topic of decentralized control theory, which provided a systematic methodology to deal with this issue. The foundations of the field were laid in the paper by Shih-Ho Wang and Edward J. Davison in the IEEE Transactions and Automatic Control, vol. 18, no. 5, in 1973, entitled "On the Stabilization of Decentralized Control Systems".

Some Highlights of the Book

- The book deals with both the theory and application of decentralized control systems.
- It includes centralized control as a special case.
- It gives some examples of real-world decentralized control case studies.
- It has a problem set at the end of each chapter (with a separate solution manual).

Toronto, Canada
Montreal, Canada
Waterloo, Canada

Edward J. Davison
Amir G. Aghdam
Daniel E. Miller

Contents

About the Authors

Edward J. Davison completed his A.R.C.T. degree in piano at the Royal Conservatory of Music in 1958 and his B.A.Sc. degree in Engineering Physics and M.A. degree in Mathematics from the University of Toronto in 1960 and 1961, respectively. He then received his Ph.D. degree and the Sc.D. degree from Cambridge University in 1964 and 1977, respectively. He is at present University Professor Emeritus of Electrical and Computer Engineering at the University of Toronto.

Dr. Davison has received several awards including the National Research Council of Canada's E.W.R. Steacie Memorial Fellowship 1974–77, the Canada Council Killam Research Fellowship 1979–80, 1981–83, the Athlone Fellowship in 1961–63 (Cambridge University), two IEEE Transactions on Automatic Control Outstanding Paper Awards and a Current Contents Classic Paper Citation. He was elected a Member of the Academy of Nonlinear Sciences (ANS), Moscow, Russia in 1998, a Fellow of the Institute of Electrical and Electronic Engineers in 1977, a Fellow of the Royal Society of Canada in 1977, an Honorary Professor of Beijing Institute of Aeronautics and Astronautics in 1986 and has been a designated Consulting Engineer of the Province of Ontario since 1979. He received the IEEE Centennial Medal in 1984, was elected a Distinguished Member of the IEEE Control Systems Society in 1984 and was awarded the International Federation of Automatic Control (IFAC) Outstanding Member Service Award in 1996. He was President of the consulting company Electrical Engineering Consociates Ltd (Toronto) from 1997 to 1999 and has served on numerous positions in the IEEE Control Systems Society, including President in 1983 and Consulting Editor of the *IEEE Transactions on Automatic Control* in 1985. He was elected to the Council of the Academy of Sciences, Royal Society of Canada, in 1999, and in 2003 he was elected a Life Fellow member of IEEE.

He was Chairman of the International Federation of Automatic Control (IFAC) Theory Committee in 1988–1990, Vice-Chairman of the IFAC Technical Board in 1990–1993, a member of the IFAC Council in 1991–1996 and a member of the IFAC Policy committee in 1996–99. He has been a member of the IFAC Administrative and Finance Committee from 1999 to 2005. He has served on numerous Editorial Boards of various journals. In 1993, he was awarded the triennial Quazza Medal from the International Federation of Automatic Control, and in December 1997, he was awarded the IEEE Control System's Hendrik W. Bode Lecture Prize. He was elected to the rank of University Professor of the University of Toronto in January 2001, and in May 2003, he was awarded the Killam Prize in Engineering from the Canada Council. In October 2003, he was inducted into the University of Toronto Engineering Alumni Hall of Distinction. He was elected a Fellow of the Canadian Academy of Engineering in June 2005, and in July 2005, he was elected a Fellow of the International Federation of Automatic Control. In 2010, he was awarded the 2010 Canada Outstanding Engineer Award given by the Institute of Electrical and Electronic Engineers (IEEE) Canada and was elected a Foreign Member of the National Academy of Engineering (NAE), USA. Dr. Davison is a Designated Consulting Engineer of the Province of Ontario, and in this capacity he has been actively involved in transfer of knowledge from academia to the industrial sector through a variety of projects.

Amir G. Aghdam received his Ph.D. in electrical and computer engineering from the University of Toronto in 2000. He is currently a Professor in the Department of Electrical and Computer Engineering at Concordia University and has served as an Associate Dean of The School of Graduate Studies (Student Affairs and Postdoctoral Studies) from 2016 to 2019. He is a member of Professional Engineers Ontario and a senior member of the IEEE. Dr. Aghdam has served as the Chair of the Conference Editorial Board of IEEE Control Systems Society, a Co-Editor-in-Chief of the IEEE Systems Journal and an Associate Editor of several journals including the IEEE Transactions on Control Systems Technology, European Journal of Control, IET Control Theory and Applications and IEEE Access. He has been a member of the Technical Program Committee of a number of conferences including IEEE Conference on Systems, Man and Cybernetics, IEEE Conference on Decision and Control and IEEE Multiconference on Systems and Control. He was a member of the Review Committee for the Italian Research and University Evaluation Agency (ANVUR) for 2012–2013 and a member of Natural Sciences and Engineering Research Council of Canada (NSERC) ECE Evaluation Group for 2014–2016. He is a recipient of the 2009 IEEE MGA Achievement Award and 2011 IEEE Canada J. J. Archambault Eastern Canada Merit Award. Dr. Aghdam was the 2014–2015 President of IEEE Canada and Director (Region 7), IEEE Inc., and was also a member of IEEE Awards Board for this period. He was a member of IEEE Medal of Honor Committee from 2017 to 2019. Dr. Aghdam was a Visiting Scholar at Harvard University in fall 2015. He has published more than 200 papers in refereed journals and conferences. His research interests include multi-agent networks, distributed control systems, sensor networks, optimization and sampled-data systems.

Daniel E. Miller received the B.Sc. degree in Electrical Engineering from the University of New Brunswick in 1984 and the M.A.Sc. and Ph.D. degrees in Electrical Engineering from the University of Toronto in 1986 and 1990, respectively. During 1990 to 1991, he held postdoctoral positions at Cambridge University, UK, the University of Groningen, The Netherlands, and the Massachusetts Institute of Technology. Since 1991 he has been a faculty member in the Department of Electrical and Computer Engineering, University of Waterloo. He has held visiting positions at the University of Alberta and the University of Newcastle, Australia, and has been an Associate Editor of the IEEE Transactions on Automatic Control. His research interests include decentralized control, adaptive control, sampled-data control, systems with delays and performance limitations.

Chapter 1
Centralized Control Systems

1.1 Introduction

When control theory is applied to a large-scale system or a multivariable system with a large number of interacting subsystems (e.g. in chemical process control, load and frequency control in power systems, transportation systems such as traffic light control or robotics with several joints), it is often desired to have some form of decentralization. In fact, for such systems, it is not realistic to assume that all output measurements can be transmitted to every local control station.

An example of a two-input, two-output system with centralized and decentralized control is given in Fig. 1.1a, b, respectively. Decentralized control techniques have similarities and differences with their centralized counterparts. It can be seen from Fig. 1.1 that the control agents of a decentralized control system have less information to use as compared to a centralized system. This chapter will be mainly focused on centralized large-scale control systems.

1.2 State-Space Model with Centralized Control

The state-space model of a finite-dimensional LTI system can be written as follows:

$$\begin{aligned} \dot{x}(t) &= Ax(t) + Bu(t) + E\omega(t) \\ y(t) &= Cx(t) + Du(t) + F\omega(t) \end{aligned} \tag{1.1}$$

where $x(t) \in R^n$ is the state vector, $u(t) \in R^m$ is the control input vector, $\omega(t) \in R^{\Omega}$ is the disturbance input, and $y(t) \in R^r$ is the output vector. Furthermore, $A \in R^{n\times n}$, $B \in R^{n\times m}$, $C \in R^{r\times n}$, $E \in R^{n\times\Omega}$ and $F \in R^{r\times\Omega}$ are constant matrices. The state-space model (1.1) represents an m-input, r-output large-scale system.

The control objectives in large-scale systems are actually similar to the objectives as in simple single-input, single-output (SISO) systems. In other words, it is

E. J. Davison et al., *Decentralized Control of Large-Scale Systems*,
https://doi.org/10.1007/978-1-4419-6014-6_1

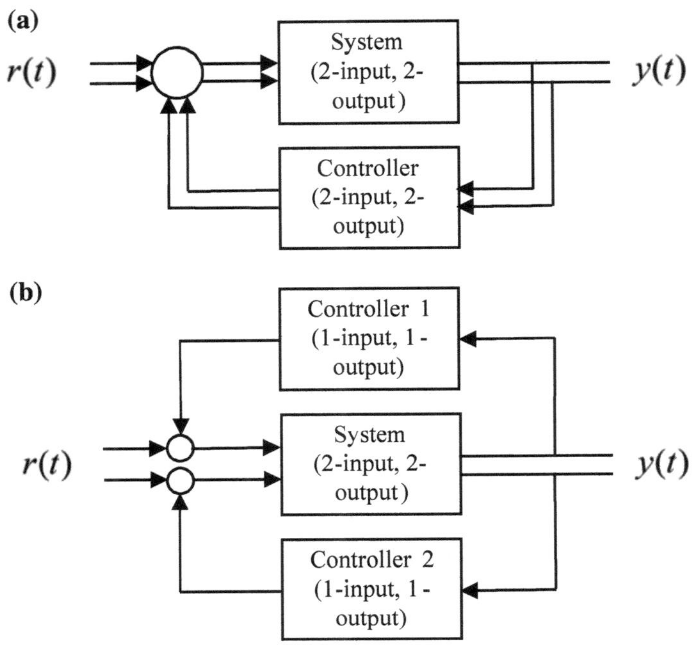

Fig. 1.1 **a** A multivariable centralized control system with two inputs and two outputs. **b** A decentralized control system with two control agents

desired to find a control law $u(t)$ such that a set of design specifications are achieved; these specifications include closed-loop stability, fast-tracking of reference signals $y_i^{\text{ref}}(t) \in R^{r_i}$ $(i = 1, \ldots, \nu)$ and rejection of disturbance signals, in a robust fashion (i.e. the tracking and disturbance rejection properties should still hold for all non-destabilizing perturbations of the plant). In addition, it may be desired to achieve these objectives in an optimal manner, by minimizing a prescribed performance index. These problems will be addressed in detail in the following chapters. However, the most important specification which is the essential prerequisite for achieving any control objective is stability, and the stability of the system (1.1) under a centralized control law will be discussed in this chapter. In this case, we will ignore any disturbances associated with plant (1.1), and so, we are concerned about the plant (1.1) with $E\omega(t) = 0$ and $F\omega(t) = 0$ given by:

$$\begin{aligned} \dot{x}(t) &= Ax(t) + Bu(t) \\ y(t) &= Cx(t) + Du(t) \end{aligned} \tag{1.2}$$

where $x(t) \in R^n$, $y(t) \in R^r$ and $u(t) \in R^m$.

It is desired now to determine if the above system can be stabilized by applying a proper LTI controller with a centralized structure; i.e., when the controller has the structure:

$$\begin{aligned} \dot{z}(t) &= Sz(t) + Ry(t) \\ u(t) &= Qz(t) \end{aligned} \tag{1.3}$$

Furthermore, it is desired to determine how "fast" the system can be stabilized. These important problems will be tackled in the sequel. To this end, the definitions of *controllability* and *observability* from classical control theory will be given next.

1.2.1 Controllability

The state of the system represented by (1.2) is said to be *controllable* if there exists a piecewise continuous input $u(t)$ that can transfer any initial state $x(t_0)$ to any final state $x(t_f)$ in a finite time interval [1].

The controllability of the system (1.2) depends only on the matrices A and B, and if the state of the system is controllable, we say that the pair (A, B) is controllable. More precisely, the system (1.2) is controllable if and only if the $n \times n.m$ *controllability matrix* is full rank [1], which is:

$$\operatorname{rank}\left(\left[B \; AB \; \ldots \; A^{n-1} \; B\right]\right) = n \tag{1.4}$$

Alternately, it is well known that the pair (A, B) is controllable if and only if:

$$\operatorname{rank}([A - \lambda_i I \;\; B]) = n, \quad i = 1, \ldots, n \tag{1.5}$$

where $\lambda_i, \; i = 1, \ldots, n$ are the eigenvalues of the matrix A. The latter criterion is called Popov–Belevitch–Hautus (PBH) test for controllability [1].

1.2.1.1 Properties of a Controllable System

It is well-known that one can freely place the modes of the system (1.2) in the complex plane using state feedback if and only if the pair (A, B) is controllable [1]. In other words, the closed-loop modes of the matrix $A + BK$ can be located arbitrarily in the s-plane using a state feedback controller $u(t) = Kx(t)$, where K is a control matrix if and only if the pair (A, B) is controllable. It is to be noted that controllability is a characteristic of the modes of the system, and using the PBH test given above, one can check the controllability of each eigenvalue of the matrix A, i.e. $\lambda_i, \; i = 1, \ldots, n$. This implies that the system (1.2) can be stabilized by using state feedback if and only if all unstable modes of the system (those eigenvalues of the matrix A which are located in closed right-half complex plane) are controllable; such a system is called *stabilizable*.

1.2.2 Observability

The state of the system represented by (1.2) is said to be *observable* if there exists a finite time T such that the initial state $x(t_0)$ can be determined from the output history $y(t),\ t_0 \leq t \leq t_0 + T$, given the input $u(t)$. Observability refers to the ability to estimate the state vector.

The observability of the system represented by (1.2) depends only on the matrices A and C, and if the state of a system with the state equations (1.2) is observable, we say that the pair (C, A) is observable. It can be shown that the system represented by (1.2) is observable if and only if the $nr \times n$ *observability matrix* is full rank, which is:

$$\operatorname{rank}\left(\begin{bmatrix} C \\ CA \\ \vdots \\ CA^{n-1} \end{bmatrix}\right) = n \tag{1.6}$$

One can also use the PBH test for observability, in a way similar to the one for controllability. In other words, the pair (C, A) is observable if and only if:

$$\operatorname{rank}\left(\begin{bmatrix} A - \lambda_i I \\ C \end{bmatrix}\right) = n, \quad i = 1, \ldots, n \tag{1.7}$$

where $\lambda_i,\ i = 1, \ldots, n$ are the eigenvalues of the matrix A.

Similar to the case of controllability, the eigenvalues of the matrix $A + LC$ can be located arbitrarily in the s-plane using an appropriate constant gain matrix L if and only if the pair (C, A) is observable. Observability is also a characteristic of the modes of the system, and using the PBH test (1.7), one can determine which modes of the system are observable. This implies that the states of the system (1.2) can be estimated by observing the output history if and only if all unstable modes of the system are observable; such a system is called *detectable*. Note that the concepts of observability and detectability are the duals of controllability and stabilizability, respectively.

Using the notions of controllability and observability, one can easily conclude that the roots of the characteristic equation of the system represented by (1.2) can be placed in any arbitrary symmetric location in the s-plane using a proper dynamic output feedback law of the form (1.3) if and only if the system is controllable and observable. Furthermore, the system can be stabilized by using a proper dynamic output feedback law if and only if the system is stabilizable and detectable. Although these results are very important in practice, it is difficult, in general, to apply the standard analytical tests to check controllability (stabilizability) and observability (detectability) of a system given by (1.4)–(1.7). This shortcoming of these standard tests will be addressed in the next section.

1.3 Centralized Fixed Modes

Consider the system (1.2), and for simplicity and without loss of generality, assume that $D = 0$. This system can be represented by the triple (C, A, B). The set of centralized fixed modes (CFM) of (C, A, B), denoted by $\Lambda(C, A, B)$, is defined as follows [2]:

$$\mathrm{CFM}(C, A, B) = \bigcap_{K \in R^{m \times r}} \mathrm{sp}(A + BKC) \tag{1.8}$$

where $\mathrm{sp}(A + BKC)$ denotes the set of eigenvalues of $(A + BKC)$.

The CFMs of the system (C, A, B) can numerically be obtained as follows:

(i) Determine $\mathrm{sp}(A) = \{\lambda_1, \ldots, \lambda_n\}$, multiplicities included.
(ii) Select an arbitrary gain matrix K so that $\|A\| \approx \|BKC\|$, where $\|.\|$ denotes the spectral norm of a matrix and is equal to the maximum singular value of the matrix. This can be accomplished by use of a pseudorandom number generator and proper scaling of the K matrix.
(iii) Then,

  (a) If $\mathrm{sp}(A+BKC)$ excludes eigenvalues $\lambda_{q1}, \lambda_{q2}, \ldots$, this implies that modes $\lambda_{q1}, \lambda_{q2}, \ldots$ are not CFMs of (C, A, B).
  (b) For almost all K [2], the set of CFMs is given by the intersection of the set of the eigenvalues of $A + BKC$.

In the SISO case, the set of CFMs is equivalent to the intersection of the set of eigenvalues of the system and the set of zeros of the system.

Remark 1.1 When an eigenvalue of A is not a CFM, but it is very close to being a CFM (i.e. for all gain matrices K selected in step (ii) of the algorithm above, one of the eigenvalues of $(A + BKC)$ is very close to one of the eigenvalues of A), it is called an approximate centralized fixed mode (ACFM). In the SISO case, an approximate CFM is a pole which is very close to a zero of the system. For example, see Fig. 1.2.

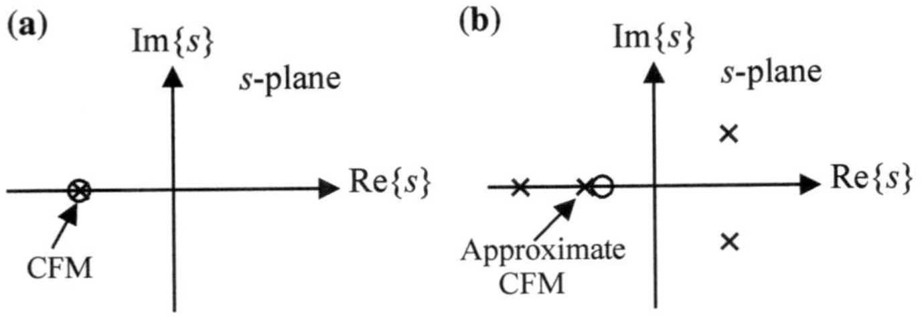

Fig. 1.2 **a** Pole-zero configuration of a SISO LTI system with a CFM; **b** pole-zero configuration of a SISO LTI system with an approximate CFM

1.4 Stability Condition for Centralized Control Systems

In order to obtain the controllability, observability, stabilizability and detectability conditions in terms of CFMs (alternative to the ones discussed in the previous subsection), note first that the definition of CFM can be used to determine certain properties of (C, A, B) as follows [2]:

- *To determine if* (A, B) *is controllable*: The pair (A, B) is controllable if and only if the triple (I_n, A, B) has no CFMs.
- *To determine if* (A, B) *is stabilizable*: The pair (A, B) is stabilizable if and only if the triple (I_n, A, B) has no CFMs which lie in the closed right-half plane of the complex plane.
- *To determine if* (C, A) *is observable*: The pair (C, A) is observable if and only if the triple (C, A, I_n) has no CFMs.
- *To determine if* (C, A) *is detectable*: The pair (C, A) is detectable if and only if the triple (C, A, I_n) has no CFMs which lie in the closed right-half plane of the complex plane.
- *To determine if* (C, A, B) *is controllable and observable*: The system represented by (C, A, B) is controllable and observable if and only if the triple (C, A, B) has no CFMs. In other words, the mode $\bar{\lambda}$ is either uncontrollable or unobservable or both if and only if it is a CFM of (C, A, B).
- *To determine if* (C, A, B) *is stabilizable and detectable*: The system represented by (C, A, B) is stabilizable and detectable if and only if the triple (C, A, B) has no CFMs which lie in the closed right-half plane of the complex plane.

The latter property implies that the system (1.2) can be stabilized by using a proper output feedback law (1.3) if and only if it does not have any CFM in the closed right-half plane. In order to determine how fast the states of the system approach their target values using a proper output feedback controller, consider the following definition.

Definition 1.1 [3] A time-varying vector $v(t)$ is said to have **a degree of exponential stability (DES)** α, if there exists a bounded scalar function $f(.)$ so that:

$$\|v(t)\| \leq f(t_0)\mathrm{e}^{-\alpha(t-t_0)}, \quad \forall t_0 \geq 0, \ t \geq t_0 \tag{1.9}$$

where α is a positive real number and $\|v(t)\|$ denotes the Euclidean norm of $v(t)$. In addition, if:

$$f(t_0) = \beta\|v(t)\|, \quad \forall t_0 \geq 0 \tag{1.10}$$

where β is a constant, then the time-varying vector $v(t)$ is said to be **uniformly stable with a DES** α.

It follows from the definition of controllability and observability, and their relationship with the notion of CFM, that the state of the system (1.2) can be stabilized with **any arbitrary degree of exponential stability** if any only if the system has no CFMs.

1.5 Minimal Realization for Centralized Systems

In many physical systems, some of the state variables in the state-space model have no effect on any of the outputs of the system. This implies that some of the state variables may be redundant and play no role in the corresponding transfer function matrix, which represents the minimal realization of the system [4]. One can use the definition of CFM to determine the minimal realization of (C, A, B) using the following algorithm:

Algorithm 1.1 [2]

(i) Find the CFMs of (C, A, B), and let them be denoted by $\{\lambda_1, \lambda_2, \ldots, \lambda_p\}$, $0 < p < n$.
(ii) Apply a gain matrix K so that the controllable and observable modes of $(C, A+BKC, B)$ are all distinct and disjoint from the CFMs $\{\lambda_1, \lambda_2, \ldots, \lambda_p\}$. This will be true for almost all K [2].
(iii) Find the eigenvalues of $(A + BKC)$, and let them be denoted by $\{\lambda_1, \lambda_2, \ldots, \lambda_p, \lambda_{p+1}, \ldots, \lambda_n\}$.
(iv) Find the eigenvectors of $(A + BKC)$ for the eigenvalues λ_i, $i = p+1, p+2, \ldots, n$, and let them be denoted by v_i, $i = p+1, p+2, \ldots, n$, respectively.
(v) Find the eigenvectors of $(A + BKC)^H$ for the eigenvalues $\tilde{\lambda}_i$, $i = p+1, p+2, \ldots, n$, where the superscript H denotes the Hermitian operator (transpose of the complex conjugate) and ~ denotes complex conjugate. Let these eigenvectors be denoted by w_i, $i = p+1, p+2, \ldots, n$, respectively.
(vi) Normalize the eigenvectors w_i, $i = p+1, p+2, \ldots, n$, so that $w_i^H.v_i = 1$, $i = p+1, p+2, \ldots, n$.
(vii) The minimal realization of $(C, A+BKC, B)$ then has the order $n-p$ and is given by $(\overline{C}, \overline{A}, \overline{B})$, where:

$$\overline{C} := C\big[v_{p+1}\ v_{p+2}\ \ldots\ v_n\big] \tag{1.11a}$$

$$\overline{A} := \text{diag}(\lambda_{p+1}, \lambda_{p+2}, \ldots, \lambda_n) \tag{1.11b}$$

$$\overline{B} := \begin{bmatrix} w_{p+1}^H \\ w_{p+2}^H \\ \vdots \\ w_n^H \end{bmatrix} B \tag{1.11c}$$

and the minimal realization of (C, A, B) is then given by $(\overline{C}, \overline{A} - \overline{B}K\overline{C}, \overline{B})$. It is to be noted that such a realization may have complex numbers; one can always convert this realization to a real form as discussed in [2].

Remark 1.2 It is to be noted that if the controllable and observable modes of (C, A, B) are all distinct and disjoint from the CFMs $\{\lambda_1, \lambda_2, \ldots, \lambda_p\}$, one can omit step (ii) and proceed to the next step by using a zero gain matrix $K = 0$. Note also that if all eigenvalues of A are distinct, one can find the vectors w_i in steps (v) and (vi) directly by forming the matrix $[v_1 \ \ldots \ v_p \ v_{p+1} \ \ldots \ v_n]$ and taking the inverse of this matrix. It can be verified that the rows of the inverse matrix are in fact the vectors w_i^H, i.e.

$$[v_1 \ \ldots \ v_p \ v_{p+1} \ \ldots \ v_n]^{-1} = \begin{bmatrix} w_1^H \\ \vdots \\ w_p^H \\ w_{p+1}^H \\ \vdots \\ w_n^H \end{bmatrix} \tag{1.12}$$

1.6 Transmission Zeros

Consider the system (1.2); the complex number λ_i is said to be a transmission zero of (1.2) if there exist an initial condition $x(0) = x_i$ and an input signal $u(t) = \begin{cases} u_i e^{\lambda_i t} & t \geq 0 \\ 0 & t < 0 \end{cases}$, $t \geq 0$, for which the output is equal to zero for all $t \geq 0$. It can be shown that the complex number λ_i is a transmission zero of (1.2) if and only if the matrix $\begin{bmatrix} A - \lambda_i I & B \\ C & D \end{bmatrix}$ is not full rank, which means that:

$$\text{rank}\left\{\begin{bmatrix} A - \lambda_i I & B \\ C & D \end{bmatrix}\right\} < n + \min(r, m). \tag{1.13}$$

The vectors x_i and u_i can be found as follows. From (1.13), it can be concluded that there exists a nonzero vector $p_i \in R^{(n+r)}$ such that:

$$\begin{bmatrix} A - \lambda_i I & B \\ C & D \end{bmatrix} p_i = 0.$$

x_i and u_i are obtained by decomposing p_i to two vectors as follows:

$$p_i = \begin{bmatrix} x_i \\ u_i \end{bmatrix}.$$

where $x_i \in R^n$ and $u_i \in R^r$. The proof is given in the appendix.

The set of transmission zeros of a system may be empty, contain a finite number of symmetric complex numbers, or include the whole complex plane. A system with a set of transmission zeros equal to the whole complex plane is called a degenerate system. Note that even a controllable and observable system with full rank matrices B and C can be degenerate. Throughout this course, we will assume that the system is non-degenerate. It can be shown from the definition of the CFM that the CFMs of a non-degenerate square system ($m = r$) are contained in the set of transmission zeros of (C, A, B) [2].

High-gain output feedback theorem [5, 6]: Consider the system (1.2), and without loss of generality, assume that $D = 0$. Let $K \in R^{m\times r}$ be an arbitrary matrix with rank $K = \min(r, m)$. Then, if $r = m$, the finite eigenvalues of $A + \rho BKC$ as $\rho \to \infty$ coincide with the transmission zeros of (C, A, B); if $r \neq m$, then for "almost all" K, the transmission zeros of (C, A, B) are contained in the finite eigenvalues of $A + \rho BKC$.

Remark 1.3 Note that the above result is a generalization of the result for SISO systems in which it is well known that for high-gain output feedback, the finite poles of the closed-loop system approach the zeros of the system.

Remark 1.4 Transmission zeros are sometimes referred to as *invariant zeros*, although the mathematical definitions of transmission zeros and invariant zeros are not exactly the same. In MATLAB, the command *tzero* can be used to find the transmission zeros of an LTI system with rational transfer function.

1.7 Model Reduction Problem

A physical system is often described by an excessively high-order model (up to hundreds of state variables). It is desired in this section to obtain a reduced-order model which approximates the behavior of the exact high-order model. This is called *the model reduction problem* [7]. Consider the system (1.2), and let the reduced-order model be represented by:

$$\begin{aligned} \dot{\bar{x}}(t) &= \overline{A}\bar{x}(t) + \overline{B}u(t) \\ y(t) &= \overline{C}\bar{x}(t) + \overline{D}u(t) \end{aligned} \tag{1.14}$$

where $\bar{x} \in R^{\bar{n}}$, $\bar{n} < n$. Usually, a large-scale system consists of a number of interconnected subsystems. Assume that the system consists of k subsystems and the order of subsystem i is equal to n_i. Then, the order of the overall system is equal to $n = \sum_{i=1}^{k} n_i$. So, if there are a large number of subsystems and each subsystem has a high order, then the overall interconnected system will have an excessively high order.

If the model (1.2) is controllable and observable, then it represents a minimal realization for the system, and any reduced-order model will only be an approximate representation for the system. The question arises as how to find the "best approximate model" for a high-order system. In order to measure the closeness of the approximate model to the original model, we need to define an appropriate performance index. This performance index can be defined based on different time responses (e.g. the step response) of the two models and can have different weightings on individual inputs and outputs and different weightings on "transient" and "steady-state" responses. It can also be defined based on the frequency response of the system. Some standard model reduction methods will be discussed in the sequel.

1.7.1 Singular Perturbation Approach

Assume that the exact model of the system can be written as:

$$\begin{aligned} \begin{bmatrix} \dot{x}_1(t) \\ \varepsilon\dot{x}_2(t) \end{bmatrix} &= \begin{bmatrix} A_1 & A_2 \\ A_3 & A_4 \end{bmatrix}\begin{bmatrix} x_1(t) \\ x_2(t) \end{bmatrix} + \begin{bmatrix} B_1 \\ B_2 \end{bmatrix}u(t) \\ y(t) &= \begin{bmatrix} C_1 & C_2 \end{bmatrix}\begin{bmatrix} x_1(t) \\ x_2(t) \end{bmatrix} + Du(t) \end{aligned} \tag{1.15}$$

where $x_1(t) \in R^{n_1}$, $x_2(t) \in R^{n-n_1}$ and ε is a "small" positive number; then, if A_4 is asymptotically stable (all eigenvalues of A_4 are in the open left-half plane), a simplified model of the exact system is given by [8]:

$$\begin{aligned} \dot{\bar{x}}(t) &= (A_1 - A_2A_4^{-1}A_3)\bar{x}(t) + (B_1 - A_2A_4^{-1}B_2)u(t) \\ y(t) &= (C_1 - C_2A_4^{-1}A_3)\bar{x}(t) + (D - C_2A_4^{-1}B_2)u(t) \end{aligned} \tag{1.16}$$

where $\bar{x}(t) \in R^{n_1}$ represents the state vector of the reduced-order model. The model (1.16) becomes "arbitrarily close" to the exact model as $\varepsilon \to 0$.

1.7.1.1 Special Case

Assume that the exact model of the system is given by:

$$\begin{bmatrix} \dot{x}_1(t) \\ \varepsilon\dot{x}_2(t) \end{bmatrix} = \begin{bmatrix} A_1 & 0 \\ 0 & A_4 \end{bmatrix}\begin{bmatrix} x_1(t) \\ x_2(t) \end{bmatrix} + \begin{bmatrix} B_1 \\ B_2 \end{bmatrix}u(t)$$
$$y(t) = \begin{bmatrix} C_1 & C_2 \end{bmatrix}\begin{bmatrix} x_1(t) \\ x_2(t) \end{bmatrix} + Du(t) \tag{1.17}$$

where $\varepsilon > 0$ and A_4 is asymptotically stable; then, a simplified model of the system is given by:

$$\dot{\bar{x}}(t) = A_1\bar{x}(t) + B_1u(t)$$
$$y(t) = C_1\bar{x}(t) + (D - C_2A_4^{-1}B_2)u(t) \tag{1.18}$$

1.7.2 Dominant Modes Approach

Assume that the pole-zero configuration of the exact model of the system is as in Fig. 1.3. In this case, if the exact system is controllable and observable, the contribution of the modes in set 1 (fast modes) in the time response of the system is negligible as $t \to \infty$, and an approximate model of the system can be obtained by deleting the modes in set 1. To clarify the idea, suppose that the exact model of the system can be diagonalized as follows:

$$\begin{bmatrix} \dot{x}_1(t) \\ \dot{x}_2(t) \end{bmatrix} = \begin{bmatrix} \Lambda_1 & 0 \\ 0 & \Lambda_2 \end{bmatrix}\begin{bmatrix} x_1(t) \\ x_2(t) \end{bmatrix} + \begin{bmatrix} B_1 \\ B_2 \end{bmatrix}u(t)$$
$$y(t) = \begin{bmatrix} C_1 & C_2 \end{bmatrix}\begin{bmatrix} x_1(t) \\ x_2(t) \end{bmatrix} + Du(t) \tag{1.19}$$

where Λ_2 is asymptotically stable. Assume also that the maximum infinity norm of $e^{\Lambda_2 t}$ is "negligible" compared to the minimum infinity norm of $e^{\Lambda_1 t}$ as time increases. In other words, the maximum diagonal element in $e^{\Lambda_2 t}$ is much smaller than the minimum diagonal element in $e^{\Lambda_1 t}$, as t increases (note that "small" is a judgment call, which depends on the specifics of the physical application). The simplified model of the system corresponding to (1.19) can then be written as follows [4, 8]:

$$\dot{\bar{x}}(t) = \Lambda_1\bar{x}(t) + B_1u(t)$$
$$y(t) = C_1\bar{x}(t) + (D - C_2\Lambda_2^{-1}B_2)u(t) \tag{1.20}$$

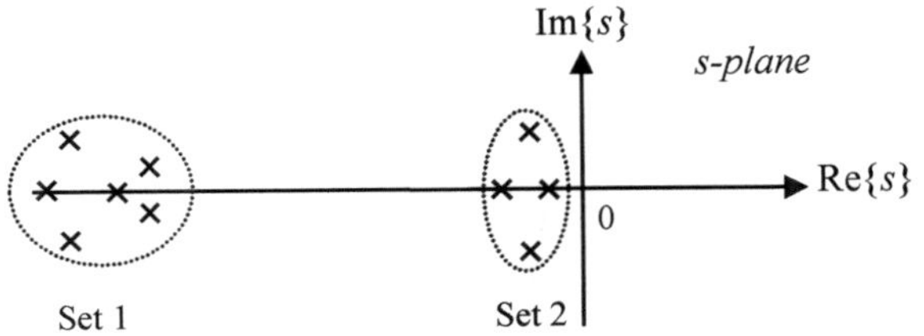

Fig. 1.3 Pole-zero configuration of a system with a set of fast poles (set 1) and a set of dominant poles (set 2)

As an example, consider the following matrices:

$$e^{\Lambda_1 t} = \begin{bmatrix} e^{-t} & 0 \\ 0 & e^{-0.1t} \end{bmatrix}, \quad e^{\Lambda_2 t} = \begin{bmatrix} e^{-100t} & 0 \\ 0 & e^{-1000t} \end{bmatrix} \tag{1.21}$$

It can probably be assumed that $e^{\Lambda_2 t}$ is negligible compared to $e^{\Lambda_1 t}$, making this system a good candidate for using the dominant modes approach to reduce the order.

The command *modreal* in MATLAB (version 7.6.0) can be used to produce a slow–fast modal realization of any given state-space model, where the set of "slow" modes and the set of "fast" modes are provided separately in one model.

1.7.3 Balanced Realization Approach

The notions of controllability and observability Gramians are essential in reducing the order of a system using the balanced realization approach. Given the system (1.2), the controllability Gramian is defined as follows:

$$W_1(0, t_f) := \int_0^{t_f} e^{A\tau} B B' e^{A'\tau} \, d\tau \tag{1.22}$$

It can be shown that this matrix is invertible at any finite time t_f, if and only if the pair (A, B) is controllable. If the system is asymptotically stable, then as $t_f \to \infty$ the matrix $W_1 := W_1(0, \infty)$ can be computed by solving the following Lyapunov equation:

$$A W_1 + W_1 A' + B B' = 0 \tag{1.23}$$

Similarly, the observability Gramian is defined as follows:

$$W_2(0, t_f) := \int_0^{t_f} e^{A'\tau} C' C e^{A\tau} \, d\tau \tag{1.24}$$

It can be shown that this matrix is invertible at any finite time t_f, if and only if the pair (C, A) is observable. If the system is asymptotically stable, then as $t_f \to \infty$, the matrix $W_2 := W_2(0, \infty)$ can be computed by solving the following Lyapunov equation:

$$A'W_2 + W_2A + C'C = 0 \tag{1.25}$$

The command *gram* in MATLAB (version 7.6.0) can be used to find the controllability and observability Gramians of an LTI system of the form (1.2).

For the physical interpretation of the controllability Gramian, assume that it is desired to transfer the state of the system (1.2) from the origin at the time $t = -T$ $(x(-T) = 0)$ to a given state x_0 at $t = 0$ $(x(0) = x_0)$, such that the input energy is minimized. In other words, it is desired to solve the following problem:

$$\text{minimize } J = \int_{-T}^{0} u'(t)u(t)\,\mathrm{d}t, \quad x(-T) = 0, \quad x(0) = x_0 \tag{1.26}$$

Note that the relationship between the input and the state is given by the state equation (1.2). It can be shown that the optimal input for the controllable system (1.2) is given by:

$$u^*(t) = B'\mathrm{e}^{-A't}[W_1(0, T)]^{-1}x_0 \tag{1.27}$$

and the corresponding energy (minimum energy) would be equal to:

$$J^* = x_0'[W_1(0, T)]^{-1}x_0 \tag{1.28}$$

If the system is asymptotically stable, then as $T \to \infty$, we will have:

$$J^* = x_0'W_1^{-1}x_0 \tag{1.29}$$

If the state x_0 is along the eigenvector of W_1 corresponding to a small eigenvalue, then J^* will be "large", which means that it would require high energy to reach this state. In other words, such states are less controllable. This implies that if W_1 is close to singularity, then the pair (A, B) is close to uncontrollability.

Similarly, for the physical interpretation of the observability Gramian, consider the system (1.2) and assume that $u(t) = 0,\ t \geq 0$. Then, the output is:

$$C\mathrm{e}^{At}x(0), \quad t \geq 0 \tag{1.30}$$

where $x(0)$ is the initial state. The output energy would be equal to:

$$\begin{aligned} E &= \int_0^T y'(t)y(t)\,\mathrm{d}t \\ &= \int_0^T x'(0)\mathrm{e}^{A't}C'C\mathrm{e}^{At}x(0)\,\mathrm{d}t \\ &= x'(0)W_2(0,T)x(0) \end{aligned} \tag{1.31}$$

If the system is asymptotically stable, then as $T \to \infty$, we will have:

$$E = x_0' W_2 x_0 \tag{1.32}$$

If the state x_0 is along the eigenvector of W_2 corresponding to a small eigenvalue, then the output energy E will be "small". This means that the effect of this initial state on the output is small. In other words, such states are less observable. This implies that if W_2 is close to singularity, then the pair (C, A) is close to being unobservable.

Note that from the controllability and observability matrices, it can only be determined if the system is controllable or observable, but from the controllability and observability Gramians, as discussed above, one can determine how much controllable and how much observable a system is. Note also that on using a similarity transformation, the controllability and observability Gramians of the system in the new coordinates will change. For example, consider the following state transformation:

$$x(t) = Tz(t) \tag{1.33}$$

Then, by substituting this transformation into (1.2), we will have:

$$\begin{aligned} \dot{z}(t) &= T^{-1}ATz(t) + T^{-1}Bu(t) := \hat{A}z(t) + \hat{B}u(t) \\ y(t) &= CTz(t) + Du(t) := \hat{C}z(t) + Du(t) \end{aligned} \tag{1.34}$$

It can be easily verified that in the new coordinates, the controllability Gramian $\hat{W}_1$ and the observability Gramian $\hat{W}_2$ will be:

$$\hat{W}_1 = T^{-1}W_1T^{-T} \tag{1.35}$$

and

$$\hat{W}_2 = T^T W_2 T \tag{1.36}$$

respectively. On the other hand, although the modes of the system under similarity transformation remain unchanged, the eigenvalues of the controllability and observability Gramians in the new coordinates may change. However, the eigenvalues of the product $\hat{W}_1 \hat{W}_2$ will be the same as $W_1 W_2$. This can be verified as follows:

$$\hat{W}_1 \hat{W}_2 = T^{-1} W_1 T^{-T} T^T W_2 T = T^{-1} W_1 W_2 T \tag{1.37}$$

Note that the eigenvalues of $T^{-1} W_1 W_2 T$ are the same as $W_1 W_2$. Note also that the eigenvalues of $W_2 W_1$ and $W_1 W_2$ are equivalent. The squared roots of the eigenvalues of $W_2 W_1$ are called the *Hankel singular values.*

It is desired now to find a new coordinate for the state equations (1.2) such that the controllability and observability Gramians are diagonal and equal, with descending diagonal terms. In other words, it is desired to have:

$$\hat{W}_1 = \hat{W}_2 = \Sigma = \begin{bmatrix} \sigma_1 & 0 & \cdots & 0 \\ 0 & \sigma_2 & \cdots & 0 \\ \vdots & \vdots & \ddots & 0 \\ 0 & 0 & 0 & \sigma_n \end{bmatrix}, \quad \sigma_1 \geq \sigma_2 \geq \cdots \geq \sigma_n > 0 \tag{1.38}$$

where σ_i, $i = 1, 2, \ldots, n$, are Hankel singular values. In such a balanced coordinate system, the controllability of each state variable is equal to the observability of that state variable. The realization $(\hat{C}, \hat{A}, \hat{B}, D)$ is called a balanced realization of the original system.

Since W_1 is positive definite ($W_1 > 0$) and symmetrical ($W_1 = W_1'$), one can find a transformation matrix V_1 such that:

$$V_1' W_1 V_1 = \Lambda_1^2 \tag{1.39}$$

where Λ_1 is a diagonal matrix with descending positive diagonal elements. Note that the columns of the matrix V_1 are the eigenvectors of the matrix W_1, with its ith column being the eigenvector of W_1 corresponding to the eigenvalue λ_i, and $\lambda_{i+1} \leq \lambda_i \leq \lambda_{i-1}$. Now, define the positive definite symmetrical matrix P as follows:

$$P = (V_1 \Lambda_1)' W_2 (V_1 \Lambda_1) \tag{1.40}$$

It can be easily verified that the eigenvalues of P are equal to those of $W_1 W_2$. Now, let the modal matrix of P be denoted by U. Then, we will have:

$$U' P U = \Sigma_p^2 = U' (V_1 \Lambda_1)' W_2 (V_1 \Lambda_1) U \tag{1.41}$$

where the ith column of U is the eigenvector of P corresponding to the eigenvalue λ_i of P, where $\lambda_{i+1} \leq \lambda_i \leq \lambda_{i-1}$. Multiply the above equation by $\Sigma^{-1/2}$ from the left and right. We will have:

$$\Sigma_p = \Sigma_p^{-1/2} (V_1 \Lambda_1 U)' W_2 (V_1 \Lambda_1 U) \Sigma_p^{-1/2} \tag{1.42}$$

Comparing the above equation with (1.36), one can conclude that by using the transformation:

$$T = V_1 \Lambda_1 U \Sigma_p^{-1/2} \tag{1.43}$$

the resulting observability Gramian will be diagonal and equal to Σ_p. It can be easily verified from (1.35) that under this transformation, the controllability Gramian will also be equal to Σ_p. In other words, we will have $\hat{W}_1 = \hat{W}_2 = \Sigma_p = \Sigma$.

The application of the balanced realization in model reduction will now be discussed. Assume that the system (1.2) is asymptotically stable and that the corresponding realization (*C*, *A*, *B*, *D*) is balanced (if it is not balanced, one can always use the above method to find the corresponding balanced realization). Now, decompose the state vector as a vector x_1 with the first n_1 state variables of x and a vector x_2 with the remaining state variables. We will have:

$$\begin{aligned}
\begin{bmatrix} \dot{x}_1(t) \\ \dot{x}_2(t) \end{bmatrix} &= \begin{bmatrix} A_1 & A_2 \\ A_3 & A_4 \end{bmatrix} \begin{bmatrix} x_1(t) \\ x_2(t) \end{bmatrix} + \begin{bmatrix} B_1 \\ B_2 \end{bmatrix} u(t) \\
y(t) &= \begin{bmatrix} C_1 & C_2 \end{bmatrix} \begin{bmatrix} x_1(t) \\ x_2(t) \end{bmatrix} + Du(t)
\end{aligned} \tag{1.44}$$

where $x_1(t) \in R^{n_1}$, $x_2(t) \in R^{n-n_1}$. Note that the controllability and observability Gramians of (1.44) are equal and can be written as:

$$\Sigma = \begin{bmatrix} \Sigma_1^{n_1 \times n_1} & 0^{n_1 \times (n-n_1)} \\ 0^{(n-n_1) \times n_1} & \Sigma_2^{(n-n_1) \times (n-n_1)} \end{bmatrix} \tag{1.45}$$

where

$$\begin{aligned}
\Sigma_1^{n_1 \times n_1} &= \begin{bmatrix} \sigma_1 & 0 & \cdots & 0 \\ 0 & \sigma_2 & \cdots & 0 \\ \vdots & \vdots & \ddots & 0 \\ 0 & 0 & 0 & \sigma_{n_1} \end{bmatrix}, \\
\Sigma_2^{(n-n_1) \times (n-n_1)} &= \begin{bmatrix} \sigma_{n_1+1} & 0 & \cdots & 0 \\ 0 & \sigma_{n_1+2} & \cdots & 0 \\ \vdots & \vdots & \ddots & 0 \\ 0 & 0 & 0 & \sigma_n \end{bmatrix}
\end{aligned} \tag{1.46}$$

Now, by removing $x_2(t)$ from the state vector, the following simplified realization:

$$\begin{aligned}
\dot{\bar{x}}(t) &= A_1 \bar{x}(t) + B_1 u(t) \\
y(t) &= C_1 \bar{x}(t) + Du(t)
\end{aligned} \tag{1.47}$$

will be a balanced realization with the controllability and observability Gramians equal to $\Sigma_1^{n_1 \times n_1}$. This can be verified by substituting (1.44) and (1.45) into the Lyapunov equations (1.23) and (1.25). Furthermore, if $\sigma_{n_1} > \sigma_{n_1+1}$ (i.e. $\sigma_{n_1} \neq \sigma_{n_1+1}$), then all the eigenvalues of A_1 will be in the open left-half plane [9], and in this case (1.47) will be an asymptotically stable, controllable and observable realization. Also, it can be shown from [9] that:

$$\begin{aligned} &\left\| \left[C(sI - A)^{-1}B + D\right] - \left[C_1(sI - A_1)^{-1}B_1 + D\right] \right\|_\infty \\ &\quad \leq 2(\sigma_{n_1+1} + \cdots + \sigma_n) = 2\text{trace}(\Sigma_2) \end{aligned} \tag{1.48}$$

This implies that the simplified realization (1.47) is a good approximation for the original asymptotically stable system (1.2), if trace(Σ_2) is sufficiently small. It can also be shown that using any approximation, the infinity norm of the error will always be greater than or equal to σ_{n_1+1} [9], i.e.

$$\left\| \left[C(sI - A)^{-1}B + D\right] - \hat{H}(s) \right\|_\infty \geq \sigma_{n_1+1}$$

where $\hat{H}(s)$ represents any proper transfer function of order n_1.

The command *balreal* in MATLAB (version 7.6.0) can be used to find the balanced realization of any state-space model.

1.8 Summary

In this chapter, some of the important issues concerning centralized linear time-invariant (LTI) finite-dimensional systems are studied. The notion of a *centralized fixed mode* (CFM) is introduced, and its relationship with the notions of controllability and observability is spelled out. In particular, a procedure is given to determine the controllability and/or observability of different modes of the system using the concept of CFM. This procedure enables one to determine the controllability and/or observability of the modes of a system, which can be computationally more efficient compared to existing analytical techniques such as the PBH test [4]. It is also shown that an LTI system can be stabilized using an LTI controller if and only if the system does not have any unstable CFMs. A highly effective algorithm is then presented to find the minimal realization of an LTI system using the concept of CFMs. The problem of model reduction for LTI systems is then studied, and different techniques are introduced. The idea behind these techniques is to retain the dominant modes of the system and delete the less dominant ones. The dominant modes may be shifted slightly in the reduced-order model in order to compensate for the neglected ones (and to maintain certain features of the system such as the DC gain). In particular, the *balanced realization* technique relies on the Gramians of controllability and observability and provides a procedure to find a similarity transformation under which the

Gramians of the controllability and observability matrices of the transformed system are diagonal and equal. An upper bound for the infinity norm of the approximation error in terms of the transfer functions of the original system and the reduced-order model is also provided, which can be used as a quantitative measure for the effectiveness of the technique in approximating a high-order model.

1.9 Problems

1. Given the system:

$$\dot{x}(t) = Ax(t) + Bu(t) + E\omega$$
$$y(t) = Cx(t) + F\omega$$

where ω is constant unmeasurable disturbance, and

$$A = \begin{bmatrix} -3 & -4 & -2 \\ 1 & 0 & 0 \\ 0 & -2 & -4 \end{bmatrix}, \quad B = \begin{bmatrix} 1 & 0 \\ 0 & 0 \\ 0 & 1 \end{bmatrix}, \quad C = \begin{bmatrix} 1 & 4 & 3 \\ 0 & 2 & 3 \end{bmatrix},$$

design a centralized controller using any method you like, to achieve "good tracking" and "good disturbance rejection" for constant reference inputs and constant disturbances, and compare the controller you obtain, with the decentralized controller:

$$u(t) = \begin{bmatrix} -40.9 & 0 \\ 0 & -42.2 \end{bmatrix} y(t) + \begin{bmatrix} -915 & 0 \\ 0 & -922 \end{bmatrix} \int_0^t (y(\tau) - y_{\mathrm{ref}}) d\tau,$$

where y_{ref} is the constant desired output.

2. Find the transmission zero(s) of the following MIMO system using the two methods discussed in this chapter.

$$\dot{x}(t) = \begin{bmatrix} 0 & 1 & 0 & 0 \\ -5 & -4 & 0.1 & 1 \\ 0.1 & 0 & -1 & 0 \\ 0 & 0 & 0 & -5 \end{bmatrix} x(t) + \begin{bmatrix} 0 & 0 \\ 0 & 0 \\ 1 & 0 \\ 0 & 1 \end{bmatrix} u(t)$$

$$y(t) = \begin{bmatrix} 16 & 4 & -4 & 2 \\ -16 & 0 & -8 & 0 \end{bmatrix} x(t)$$

3. Consider the following system:

$$\begin{aligned}\dot{x}_1(t) &= -3x_1(t) + x_2(t) - x_3(t) + u(t)\\ \dot{x}_2(t) &= -x_3(t) + u(t)\\ \dot{x}_3(t) &= x_1(t) - ax_3(t)\\ y(t) &= x_1(t)\end{aligned}$$

where $x_i(t)$ and $i = 1, 2, 3$ are the state variables, $u(t)$ is the control input, and $y(t)$ is the output.

(a) Write the state equations in the standard matrix form. Write a MATLAB code and change the value of a between 0.1 and 2. Check for what value of a the system has CFM.
(b) Check the controllability and observability of the system for the value of a obtained in part (a).
(c) Find the minimal realization of the system for the value of a obtained in part (a).

4. A model of the temperature control system for a building is as follows:

$$\begin{aligned}\dot{x}(t) &= Ax(t) + Bu(t)\\ y(t) &= Cx(t)\end{aligned}$$

with $x(t) \in R^{20}$, $u(t) \in R^4$ and $y(t) \in R^4$ given in the CD.

(a) Determine if the system has any CFM.
(b) Determine if the system is controllable.
(c) Determine if the system is observable.
(d) If the system is not controllable and observable, find a minimal realization of the system.

5. It is desired to obtain an approximate minimal realization (reduced model) of order 4 for the temperature control system given in Problem 4. Making suitable assumptions, find an LTI model of order 4, which approximates the behavior of the original system. Compare the step response of the approximate model obtained with the step response of the original system.

6. Given an asymptotically stable finite-dimensional LTI system, show that the DC gain of the reduced-order model obtained by using a singular perturbation approach is the same as that of the original system.

7. Consider the following controllable and observable system:

$$\begin{aligned}\dot{x}(t) &= Ax(t) + Bu(t)\\ y(t) &= Cx(t) + Du(t)\end{aligned}$$

Using a procedure similar to the one for obtaining a balanced realization, find a transformation matrix T to change the coordinates of the system such that the controllability and observability Gramians in the new coordinate system are:

(a) $\hat{W}_2 = \Sigma^2,\ \hat{W}_1 = I$ (the corresponding realization is called input normal).
(b) $\hat{W}_2 = \Sigma^2,\ \hat{W}_1 = \Sigma^2$ (the corresponding realization is called output normal).

8. Consider the following SISO system:

$$\dot{x}(t) = \begin{bmatrix} -31 & -430 & -2400 & -2000 \\ 1 & 0 & 0 & 0 \\ 0 & 1 & 0 & 0 \\ 0 & 0 & 1 & 0 \end{bmatrix} x(t) + \begin{bmatrix} 1 \\ 0 \\ 0 \\ 0 \end{bmatrix} u(t)$$
$$y(t) = \begin{bmatrix} 0 & 0 & \frac{20{,}000}{3} & 2000 \end{bmatrix} x(t)$$

(a) Find transformation matrix T to obtain a balanced realization for this system.
(b) Find the balanced realization for this system.
(c) Use the balanced realization to obtain a third-order approximate model for the system.
(d) What is the maximum infinity norm of the error (the difference between the exact model and the approximate model)?
(e) Compare the Bode plots (magnitude and phase) of the exact model and the approximate model.

9. A large flexible space structure may, under certain conditions, be described by the LTI system:

$$\dot{x}(t) = \begin{bmatrix} 0_{n\times n} & I_n \\ -\Omega^2 & 0_{n\times n} \end{bmatrix} x(t) + \begin{bmatrix} 0_{n\times m} \\ B \end{bmatrix} u(t);\ x(t) \in R^{2n},\ u(t) \in R^m$$
$$y(t) = \begin{bmatrix} B' & 0 \end{bmatrix} x(t);\ y(t) \in R^m$$

where

$$-\Omega^2 := \mathrm{diag}([\ \overbrace{0 \cdots 0}^{r \text{ "rigid modes"}} \quad \overbrace{-\omega_1^2 \cdots -\omega_p^2}^{p \text{ "static modes"}}]);\ r + p = n,\ m \geq r$$

I_n is the $n \times n$ identity matrix, and $B \in R^{n\times m}$. What can be said about the transmission zeros of the system?

10. Consider the following system:

$$\dot{x}(t) = Ax(t) + Bu(t)$$
$$y(t) = Cx(t)$$

Find a similarity transformation under which $\overline{W}_1$, the controllability Gramian in the new coordinate, is equal to $k \times I$, where I is the identity matrix with a proper dimension and k is any given positive scalar.

References

1. K. Ogata, *Modern Control Engineering*, 4th edn. (Prentice Hall, New Jersey, 2002)
2. E.J. Davison, W. Gesing, S.H. Wang, An algorithm for obtaining the minimal realization of a linear time-invariant system and determining if a system is stabilizable-detectable. IEEE Trans. Autom. Control **AC-23**, 1048–1054 (1978)
3. Z. Gong, M. Aldeen, Stabilization of decentralized control systems. J. Math. Syst. Estimation Control **7**(1), 1–16 (1997)
4. W. Brogan, *Modern Control Theory*, 3rd edn. (Prentice Hall, New Jersey, 1991)
5. E.J. Davison, S.H. Wang, An algorithm for calculation of transmission zeros of the system (C, A, B, D) using high gain output feedback. IEEE Trans. Autom. Control **AC-23**, 738–741 (1978)
6. E.J. Davison, S.H. Wang, Properties and calculation of transmission zeros of linear multivariable systems. Automatica **10**, 643–658 (1974)
7. B. Moore, Principal component analysis in linear systems: controllability, observability, and model reduction. IEEE Trans. Autom. Control **AC-26**(1), 17–31 (1981)
8. E.J. Davison, A method for simplifying linear dynamic systems. IEEE Trans. Autom. Control **AC-11**(1), 93–101 (1966)
9. K. Glover, All optimal Hankel-norm approximation of linear multivariable systems and their $L, ^{\infty}$ error bounds. Int. J. Control **39**(6), 1115–1193 (1984)
10. E.J. Davison, A. Solomon, Partial decentralized temperature control of multi-zone buildings, in *Proceedings of the 22nd IEEE Conference on Decision and Control* (1983), pp. 10–16

Chapter 2
Decentralized Control Systems

2.1 State-Space Representation of Decentralized Control Systems

Consider an LTI system with ν local control stations, which is represented by:

$$\begin{aligned}\dot{x}(t) &= Ax(t) + \begin{bmatrix} B_1 & \cdots & B_\nu \end{bmatrix} \begin{bmatrix} u_1(t) \\ \vdots \\ u_\nu(t) \end{bmatrix} \\ \begin{bmatrix} y_1(t) \\ \vdots \\ y_\nu(t) \end{bmatrix} &= \begin{bmatrix} C_1 \\ \vdots \\ C_\nu \end{bmatrix} x(t) + \begin{bmatrix} D_{11} & \cdots & D_{1\nu} \\ \vdots & \ddots & \vdots \\ D_{\nu 1} & \cdots & D_{\nu\nu} \end{bmatrix} \begin{bmatrix} u_1(t) \\ \vdots \\ u_\nu(t) \end{bmatrix}\end{aligned} \tag{2.1}$$

where $x(t) \in R^n$ is the state vector, $u_i(t) \in R^{m_i}$ and $y_i(t) \in R^{r_i}$ are the input and output, respectively, of the ith control station ($i = 1, \ldots, \nu$). The matrices $A \in R^{n \times n}$, $B_i \in R^{n \times m_i}$, $C_i \in R^{r_i \times n}$ and $D_{ij} \in R^{r_i \times m_j}$ ($i, j = 1, \ldots, \nu$) are real, constant matrices. The state-space model given by (2.1) presents the parameters of each control agent explicitly and hence is more appropriate [as compared to the model (1.2)] for representing decentralized control systems. It is sometimes more convenient to rewrite (2.1) in the following form:

$$\begin{aligned}\dot{x}(t) &= Ax(t) + \sum_{i=1}^{\nu} B_i u_i(t) \\ y_i(t) &= C_i x(t) + \sum_{j=1}^{\nu} D_{ij} u_j(t), \quad i = 1, \ldots, \nu\end{aligned} \tag{2.2}$$

E. J. Davison et al., *Decentralized Control of Large-Scale Systems*,
https://doi.org/10.1007/978-1-4419-6014-6_2

The set of local dynamic LTI feedback controllers for (2.2) are given by:

$$
\begin{aligned}
\dot{z}_i(t) &= S_i z_i(t) + R_i y_i(t) \\
u_i(t) &= Q_i z_i(t) + K_i y_i(t) + v_i(t), \quad i = 1, \ldots, \nu
\end{aligned}
\tag{2.3}
$$

where $z_i(t) \in R^{\eta_i}$ is the state vector of the ith feedback controller, and $v_i(t) \in R^{m_i}$ is the ith local external input. The matrices $S_i \in R^{\eta_i \times \eta_i}$, $R_i \in R^{\eta_i \times m_i}$, $Q_i \in R^{r_i \times \eta_i}$ and $K_i \in R^{r_i \times m_i}$ $(i = 1, \ldots, \nu)$ are real, constant matrices. The controller (2.3) can be written in the following form:

$$
\begin{aligned}
\dot{z}(t) &= Sz(t) + Ry(t) \\
u(t) &= Qz(t) + Ky(t) + v(t)
\end{aligned}
\tag{2.4}
$$

where S, R, Q and K are block diagonal matrices as follows:

$$
S := \begin{bmatrix} S_1 & & \mathbf{0} \\ & \ddots & \\ \mathbf{0} & & S_\nu \end{bmatrix}, R := \begin{bmatrix} R_1 & & \mathbf{0} \\ & \ddots & \\ \mathbf{0} & & R_\nu \end{bmatrix}, Q := \begin{bmatrix} Q_1 & & \mathbf{0} \\ & \ddots & \\ \mathbf{0} & & Q_\nu \end{bmatrix}, K := \begin{bmatrix} K_1 & & \mathbf{0} \\ & \ddots & \\ \mathbf{0} & & K_\nu \end{bmatrix}, \tag{2.5}
$$

and $z(t)$, $y(t)$, $u(t)$ and $v(t)$ are given by:

$$
z(t) := \begin{bmatrix} z_1(t) \\ \vdots \\ z_\nu(t) \end{bmatrix}, y(t) := \begin{bmatrix} y_1(t) \\ \vdots \\ y_\nu(t) \end{bmatrix}, u(t) := \begin{bmatrix} u_1(t) \\ \vdots \\ u_\nu(t) \end{bmatrix}, v(t) := \begin{bmatrix} v_1(t) \\ \vdots \\ v_\nu(t) \end{bmatrix} \tag{2.6}
$$

Note that in the s-domain, the controller (2.4) will have the following form:

$$
\begin{bmatrix} U_1(s) \\ \vdots \\ U_\nu(s) \end{bmatrix} = \begin{bmatrix} G_{C1}(s) & & \mathbf{0} \\ & \ddots & \\ \mathbf{0} & & G_{C\nu} \end{bmatrix} \begin{bmatrix} Y_1(s) \\ \vdots \\ Y_\nu(s) \end{bmatrix} + V(s) \tag{2.7}
$$

Using the augmented state vector $\begin{bmatrix} x(t) \\ z(t) \end{bmatrix}$, one can write the equations for the closed-loop system as follows:

$$
\begin{aligned}
\begin{bmatrix} \dot{x}(t) \\ \dot{z}(t) \end{bmatrix} &= \overline{A} \begin{bmatrix} x(t) \\ z(t) \end{bmatrix} + \overline{B} v(t) \\
y(t) &= \overline{C} \begin{bmatrix} x(t) \\ z(t) \end{bmatrix} + \overline{D} v(t)
\end{aligned}
\tag{2.8}
$$

For a strictly proper system ($D = 0$), the matrices $\overline{A}$, $\overline{B}$ and $\overline{C}$ are given by:

$$\overline{A} := \begin{bmatrix} A + BKC & BQ \\ RC & S \end{bmatrix}, \overline{B} := \begin{bmatrix} B \\ 0 \end{bmatrix}, \overline{C} := \begin{bmatrix} C & 0 \end{bmatrix}, \overline{D} := 0 \tag{2.9}$$

where

$$B := \begin{bmatrix} B_1 & \cdots & B_\nu \end{bmatrix}, C := \begin{bmatrix} C_1 \\ \vdots \\ C_\nu \end{bmatrix} \tag{2.10}$$

Note that the equations for a centralized controller are also similar to (2.4). However, for a decentralized controller, the matrices S, R, Q and K are block diagonal as given by (2.5), whereas in the centralized case (Eq. (1.3)), there is no such restriction on the structure of these matrices.

2.2 Decentralized Fixed Modes

Consider the m-input, r-output system (2.1), where $m = \sum_{i=1}^{\nu} m_i$, $r = \sum_{i=1}^{\nu} r_i$, and assume that the decentralized flow constraint $\mathbf{K}$ is defined as follows:

$$\mathbf{K} := \{K \in R^{m\times r} | K = \begin{bmatrix} K_1 & & \mathbf{0} \\ & \ddots & \\ \mathbf{0} & & K_\nu \end{bmatrix}, K_i \in R^{m_i \times r_i}, i = 1, \ldots, \nu, \det(I - DK) \neq 0\} \tag{2.11}$$

Then, $\lambda \in C$ is said to be a *decentralized fixed mode* (DFM) of (2.1) with respect to $\mathbf{K}$, if:

$$\lambda \in \bigcap_{\forall K \in \mathbf{K}} \mathrm{sp}(A + BK(I - DK)^{-1}C) \tag{2.12}$$

where $\mathrm{sp}(A + BK(I - DK)^{-1}C)$ denotes the set of eigenvalues of $(A + BK(I - DK)^{-1}C)$. In other words, $\lambda \in C$ is a DFM of (2.1) with respect to $\mathbf{K}$, if:

$$\mathrm{rank}(A - \lambda I + BK(I - DK)^{-1}C) < n, \quad \forall K \in \mathbf{K} \tag{2.13}$$

For strictly proper systems, Eq. (2.12) can be simplified as:

$$\lambda \in \bigcap_{\forall K \in \mathbf{K}} \mathrm{sp}(A + BKC) \tag{2.14}$$

This is, in fact, equivalent to the following:

$$\lambda \in \bigcap_{\forall K_i \in R^{m_i \times r_i}} \mathrm{sp}\left(A + \sum_{i=1}^{\nu} B_i K_i C_i\right) \tag{2.15}$$

Equations (2.14) and (2.15) can be used to find the DFMs of a system numerically. Note that the set of CFMs of any LTI MIMO system is a subset of the set of its DFMs.

2.2.1 Invariance of Decentralized Fixed Modes

Consider the system (2.1) with the decentralized dynamic controller (2.4). A mode $\lambda \in \mathrm{sp}(A)$ is a DFM of system (2.1) if and only if for all LTI controllers of the form (2.4), λ is an eigenvalue of the closed-loop system matrix of (2.8). The following numerical algorithm can be used to determine DFMs of (2.1) [2].

Algorithm 2.1

(1) Find the eigenvalues of A.
(2) Select an arbitrary block diagonal feedback gain matrix $K \in \mathbf{K}$ such that the matrix $(I - DK)$ is non-singular. This can be accomplished by use of a pseudorandom number generator (numerically, it is better to properly scale the gain matrix K such that $\|A\| \approx \left\|BK(I - DK)^{-1}C\right\|$).
(3) Find the eigenvalues of the matrix $A_c := A + BK(I - DK)^{-1}C$.
(4) For almost all $K \in \mathbf{K}$, the set of DFMs with respect to $\mathbf{K}$ is equal to the intersection set $\mathrm{sp}(A) \bigcap_{\forall K \in \mathbf{K}} \mathrm{sp}(A_c)$.

2.2.2 Stability Conditions for Decentralized Systems

The following theorem provides necessary and sufficient conditions for the stabilizability of LTI systems under decentralized LTI output feedback.

Theorem 2.1 [1, 2] *Consider the system (C, A, B, D) given by (2.1), with B and C defined in (2.10). Let $\mathbf{K}$ be the set of block diagonal matrices defined in (2.11). Then, a necessary and sufficient condition for the existence of a decentralized LTI controller given by (2.4) such that the closed-loop system is asymptotically stable is that the system has no DFMs in the closed right-half complex plane.*

Example 2.1 Consider the system (2.1) with $\nu = 2$ and the following system parameters:

$$A = \begin{bmatrix} -1 & 0 & -3 \\ 0 & \alpha & 0 \\ 0 & 0 & -3 \end{bmatrix}, B_1 = \begin{bmatrix} 1 \\ 0 \\ 1 \end{bmatrix}, B_2 = \begin{bmatrix} 0 \\ 1 \\ 1 \end{bmatrix}, C_1 = \begin{bmatrix} 0 & 1 & 0 \end{bmatrix}, C_2 = \begin{bmatrix} -1.1 & 0 & 0.1 \end{bmatrix}, D = \begin{bmatrix} 0 & 0 \\ 0 & 0 \end{bmatrix} \tag{2.16}$$

It can be easily seen that this system is controllable and observable, and so, it has no CFMs. However, we have:

$$A + B\begin{bmatrix} K_1 & 0 \\ 0 & K_2 \end{bmatrix}C = \begin{bmatrix} -1 & K_1 & -3 \\ -1.1K_2 & \alpha & 0.1K_2 \\ -1.1K_2 & K_1 & -3+0.1K_2 \end{bmatrix} \tag{2.17}$$

One can verify that for $\alpha = 0.1$, this system has a DFM at $\lambda = 0.1$ with respect to the diagonal information flow $\begin{bmatrix} \times & 0 \\ 0 & \times \end{bmatrix}$. Hence, it follows from Theorem 2.1 that the system cannot be stabilized by using a decentralized LTI controller.

2.3 Characterization of Decentralized Fixed Modes

Two main analytical methods will be introduced in the sequel to determine DFMs of (2.1).

2.3.1 PBH-Type Criteria

As was pointed out in Chap. 4, a finite-dimensional LTI system is controllable and observable if and only if it has no CFMs. One can then use Popov–Belevitch–Hautus (PBH) test to find the CFMs. In the next theorem, PBH-type criteria are provided to find the DFMs of a general proper finite-dimensional LTI system.

Theorem 2.2 [2] *Consider the system given by (2.1). Then,* $\lambda \in \mathrm{sp}(A)$ *is a DFM of the system with respect to* $\mathbf{K}$ *if and only if the following PBH-type condition holds:*

$$\mathrm{rank}\left(\begin{bmatrix} A-\lambda I & B_{i_1} & \dots & B_{i_p} \\ C_{i_{p+1}} & D_{i_{p+1},i_1} & \dots & D_{i_{p+1},i_p} \\ \vdots & \vdots & & \vdots \\ C_{i_\nu} & D_{i_\nu,i_1} & \dots & D_{i_\nu,i_p} \end{bmatrix}\right) < n \tag{2.18}$$

for some distinct integers $i_1, i_2, \dots, i_\nu$ *such that* $\{i_1, i_2, \dots, i_\nu\} = \{1, 2, \dots, \nu\}$, *and some* $p \in \{1, 2, \dots, \nu\}$.

Remark 2.1 It is to be noted that for $p = \nu$, (2.18) is the condition for uncontrollability. Similarly, for $p = \phi$, it is the condition for unobservability. In other words, if the condition (2.18) of Theorem 2.2 is satisfied for $p = \nu$ or $p = \phi$, then the corresponding DFM is also a CFM of the system (2.1).

Example 2.2 Consider the system given in Example 2.1. We want to use Theorem 2.2 to find the value of α for which the system has a DFM. We will check the rank of $\nu + 1 = 3$ matrices in Theorem 2.2.
The matrix corresponding to Step 1 of Theorem 2.2:

$$M_1 = \begin{bmatrix} A - \lambda I \\ C_1 \\ C_2 \end{bmatrix} \tag{2.19}$$

The matrix corresponding to Step 2 of Theorem 2.2:

$$M_2 = \begin{bmatrix} A - \lambda I & B_1 & B_2 \end{bmatrix} \tag{2.20}$$

The matrices corresponding to Step 3 of Theorem 2.2:

$$M_3 = \begin{bmatrix} A - \lambda I & B_2 \\ C_1 & 0 \end{bmatrix}, M_4 = \begin{bmatrix} A - \lambda I & B_1 \\ C_2 & 0 \end{bmatrix} \tag{2.21}$$

It can be verified that for all eigenvalues of A ($\lambda \in \{-1, \alpha, -3\}$) the matrices M_1 and M_2 are full rank. In other words, the system is observable and controllable. However, the rank of the matrix M_4 for $\lambda = \alpha = 0.1$ will be less than 3, as follows:

$$\text{rank}(M_4) = \text{rank}\left(\begin{bmatrix} -1.1 & 0 & -3 & 1 \\ 0 & 0 & 0 & 0 \\ 0 & 0 & -3.1 & 1 \\ -1.1 & 0 & 0.1 & 0 \end{bmatrix}\right) = 2 < n = 3 \tag{2.22}$$

From Theorem 2.2, it will be concluded that the system has a DFM at $\lambda = 0.1$ for $\alpha = 0.1$.

We will now discuss the second method to analytically obtain the DFMs of a system.

2.3.2 *Transmission Zero Criteria*

2.3.2.1 Single-Input, Single-Output (SISO) Subsystems

Theorem 2.3 [2] *Consider the system given by (2.1), and assume for simplicity that the system is strictly proper ($D_{ij} = 0$, for all $i, j = 1, \ldots, \nu$) and that all control stations are SISO ($m_i = r_i = 1$, for all $i = 1, \ldots, \nu$). Let the matrices B_i and C_i be denoted by b_i and c_i, respectively, to emphasize the fact that they are column and row vectors. Then, $\lambda \in \mathrm{sp}(A)$ is a DFM of (2.1) with respect to the diagonal information flow **K** (the set of all $\nu \times \nu$ diagonal matrices) if and only if λ is a transmission zero of all of the following systems:*

(1) (c_i, A, b_i),
$i = 1, 2, \ldots, \nu$

(2) $\left(\begin{bmatrix} c_i \\ c_j \end{bmatrix}, A, \begin{bmatrix} b_i & b_j \end{bmatrix}\right)$,
$i = 1, 2, \ldots, \nu - 1;\ j = i + 1, i + 2, \ldots, \nu$

(3) $\left(\begin{bmatrix} c_i \\ c_j \\ c_k \end{bmatrix}, A, \begin{bmatrix} b_i & b_j & b_k \end{bmatrix}\right)$,
$i = 1, 2, \ldots, \nu-2;\ j = i+1, i+2, \ldots, \nu-1;\ k = j+1, j+2, \ldots, \nu$

$\vdots$

(v) $\left(\begin{bmatrix} c_1 \\ c_2 \\ \vdots \\ c_\nu \end{bmatrix}, A, \begin{bmatrix} b_1 & b_2 & \cdots & b_\nu \end{bmatrix}\right)$.

Example 2.3 Consider the system (2.1) with $\nu = 3$ and the system parameters $A \in R^{n\times n}$, $B = \begin{bmatrix} b_1 & b_2 & b_3 \end{bmatrix}$, $b_1, b_2, b_3 \in R^{n\times 1}$, $C = \begin{bmatrix} c_1 \\ c_2 \\ c_3 \end{bmatrix}$, $c_1, c_2, c_3 \in R^{1\times n}$ and $D = 0$. Then, $\lambda \in \mathrm{sp}(A)$ is a DFM of this system with respect to the 3×3 diagonal information flow **K** if and only if all of the following inequalities hold:

(5) $\mathrm{rank}\left(\begin{bmatrix} A - \lambda I & b_1 \\ c_1 & 0 \end{bmatrix}\right) < n + 1,$

$\mathrm{rank}\left(\begin{bmatrix} A - \lambda I & b_2 \\ c_2 & 0 \end{bmatrix}\right) < n + 1,$

$$\text{rank}\left(\begin{bmatrix} A-\lambda I & b_3 \\ c_3 & 0 \end{bmatrix}\right) < n+1,$$

$$(6)\quad \text{rank}\left(\begin{bmatrix} A-\lambda I & b_1 & b_2 \\ c_1 & 0 & 0 \\ c_2 & 0 & 0 \end{bmatrix}\right) < n+2,$$

$$\text{rank}\left(\begin{bmatrix} A-\lambda I & b_1 & b_3 \\ c_1 & 0 & 0 \\ c_3 & 0 & 0 \end{bmatrix}\right) < n+2,$$

$$\text{rank}\left(\begin{bmatrix} A-\lambda I & b_2 & b_3 \\ c_2 & 0 & 0 \\ c_3 & 0 & 0 \end{bmatrix}\right) < n+2,$$

$$(7)\quad \text{rank}\left(\begin{bmatrix} A-\lambda I & b_1 & b_2 & b_3 \\ c_1 & 0 & 0 & 0 \\ c_2 & 0 & 0 & 0 \\ c_3 & 0 & 0 & 0 \end{bmatrix}\right) < n+3.$$

2.3.2.2 Multi-input, Multi-output (MIMO) Subsystems

Consider now a more general case of the system (2.1), where the system is strictly proper, but each control station is not necessarily SISO. We have:

$$B := \begin{bmatrix} B_1 & \cdots & B_\nu \end{bmatrix} = \begin{bmatrix} b_{11} & \cdots & b_{1m_1} & \cdots & b_{\nu 1} & \cdots & b_{\nu m_\nu} \end{bmatrix},$$

$$C := \begin{bmatrix} C_1 \\ \vdots \\ C_\nu \end{bmatrix} = \begin{bmatrix} c_{11} \\ \vdots \\ c_{1r_1} \\ \vdots \\ c_{\nu 1} \\ \vdots \\ c_{\nu r_\nu} \end{bmatrix},$$

$$\mathbf{K} := \left\{ K \in R^{m\times r} \,|\, K = \begin{bmatrix} K_1 & & \mathbf{0} \\ & \ddots & \\ \mathbf{0} & & K_\nu \end{bmatrix} = \begin{bmatrix} \begin{matrix} k_{1,\,11} & \cdots & k_{1,\,1r_1} \\ \vdots & \ddots & \vdots \\ k_{1,\,m_1 1} & \cdots & k_{1,\,m_1 r_1} \end{matrix} & & \mathbf{0} \\ & \ddots & \\ \mathbf{0} & & \begin{matrix} k_{\nu,\,11} & \cdots & k_{\nu,\,1r_\nu} \\ \vdots & \ddots & \vdots \\ k_{\nu,\,m_\nu 1} & \cdots & k_{\nu,\,m_\nu r_\nu} \end{matrix} \end{bmatrix} \right\}.$$

By expanding $A + BKC$ in terms of the rows of C and columns of B, we will have:

$$\begin{aligned} A + BKC = A &+ b_{11}k_{1,\,11}c_{11} + \cdots + b_{1m_1}k_{1,\,m_1 1}c_{11} + \cdots + b_{11}k_{1,\,1r_1}c_{1r_1} \\ &+ \cdots + b_{1m_1}k_{1,\,m_1 r_1}c_{1r_1} + b_{\nu 1}k_{\nu,\,11}c_{\nu 1} + \cdots + b_{\nu m_\nu}k_{\nu,\,m_\nu 1}c_{\nu 1} \\ &+ \cdots + b_{\nu 1}k_{\nu,\,1r_\nu}c_{\nu r_\nu} + \cdots + b_{\nu m_\nu}k_{\nu,\,m_\nu r_\nu}c_{\nu r_\nu} \end{aligned}$$

Note that $k_{1,11}, \ldots, k_{\nu,m_\nu r_\nu}$ are all scalars. This implies that (C, A, B) with the block diagonal decentralized information flow $\mathbf{K}$ is equivalent to $(\overline{C}, A, \overline{B})$ with diagonal information flow $\overline{\mathbf{K}}$, where

$$\overline{C} = \begin{bmatrix} \left.\begin{matrix} c_{11} \\ \vdots \\ c_{11} \end{matrix}\right\} m_1 \text{ rows} \\ \vdots \\ \left.\begin{matrix} c_{1r_1} \\ \vdots \\ c_{1r_1} \end{matrix}\right\} m_1 \text{ rows} \\ \vdots \\ \left.\begin{matrix} c_{\nu 1} \\ \vdots \\ c_{\nu 1} \end{matrix}\right\} m_\nu \text{ rows} \\ \vdots \\ \left.\begin{matrix} c_{\nu r_\nu} \\ \vdots \\ c_{\nu r_\nu} \end{matrix}\right\} m_\nu \text{ rows} \end{bmatrix} \qquad \begin{matrix} \text{(first two groups: } r_1 \text{ blocks)} \\ \text{(last two groups: } r_\nu \text{ blocks)} \end{matrix}$$

$$\overline{B} = \Big[\underbrace{\underbrace{b_{11} \;\cdots\; b_{1m_1}}_{m_1 \text{ columns}} \;\cdots\; \underbrace{b_{11} \;\cdots\; b_{1m_1}}_{m_1 \text{ columns}}}_{r_1 \text{ blocks}} \;\cdots\; \underbrace{\underbrace{b_{\nu 1} \;\cdots\; b_{\nu m_\nu}}_{m_\nu \text{ columns}} \;\cdots\; \underbrace{b_{\nu 1} \;\cdots\; b_{\nu m_\nu}}_{m_\nu \text{ columns}}}_{r_\nu \text{ blocks}}\Big]$$

and $\overline{\mathbf{K}}$ is the set of all diagonal matrices $\overline{K}$ of the following form:

$$
\overline{K} = \left[\begin{array}{ccccccc}
\overbrace{\overbrace{\begin{bmatrix} k_{1,11} & & \mathbf{0} \\ & \ddots & \\ \mathbf{0} & & k_{1,m_1 1} \end{bmatrix}}^{m_1 \text{ columns}} & \ddots & \overbrace{\begin{bmatrix} k_{1,1r_1} & & \mathbf{0} \\ & \ddots & \\ \mathbf{0} & & k_{1,m_1 r_1} \end{bmatrix}}^{m_1 \text{ columns}}}^{r_1 \text{ blocks}} & & & & \mathbf{0} \\
& & & \ddots & & & \\
& & & & \overbrace{\overbrace{\begin{bmatrix} k_{\nu,11} & & \mathbf{0} \\ & \ddots & \\ \mathbf{0} & & k_{\nu,m_\nu 1} \end{bmatrix}}^{m_\nu \text{ columns}} & \ddots & \overbrace{\begin{bmatrix} k_{\nu,1r_\nu} & & \mathbf{0} \\ & \ddots & \\ \mathbf{0} & & k_{\nu,m_\nu r_\nu} \end{bmatrix}}^{m_\nu \text{ columns}}}^{r_\nu \text{ blocks}} \\
\mathbf{0} & & & & & &
\end{array}\right]
$$

This means that $\lambda \in \mathrm{sp}(A)$ is a DFM of the system (C, A, B) with respect to the block diagonal information flow $\mathbf{K}$ if it is a DFM of the system $(\overline{C}, A, \overline{B})$ with respect to the diagonal information flow $\overline{\mathbf{K}}$. Note that the matrices $\overline{C}$ and $\overline{B}$ can be obtained from C and B by using the Kronecker product (this will be shown in the next few pages).

Consider now the system (2.1) in the most general case (i.e. a general proper system with MIMO control agents). This system can be written in the following form:

$$
\begin{aligned}
\dot{x}(t) &= Ax(t) + \sum_{i=1}^{\nu^*} b_i u_i^*(t) \\
y_i^*(t) &= c_i x(t) + \sum_{j=1}^{\nu^*} d_{ij} u_j^*(t), \quad i = 1, \ldots, \nu^*
\end{aligned}
\tag{2.23}
$$

where $u_i^*(t)$ are the scalar inputs, $y_i^*(t)$ are the scalar outputs, $\nu^* = \sum_{i=1}^{\nu} m_i r_i$, and where

$$\begin{bmatrix} b_1 & \cdots & b_{\nu*} \end{bmatrix} := \begin{bmatrix} \alpha(r_1) \otimes B_1 & \cdots & \alpha(r_\nu) \otimes B_\nu \end{bmatrix}$$

$$\begin{bmatrix} c_1 \\ \vdots \\ c_{\nu*} \end{bmatrix} := \begin{bmatrix} C_1 \otimes \alpha(m_1)' \\ \vdots \\ C_\nu \otimes \alpha(m_\nu)' \end{bmatrix}$$

$$\begin{bmatrix} d_{11} & \cdots & d_{1\nu*} \\ \vdots & \ddots & \vdots \\ d_{\nu*1} & \cdots & d_{\nu*\nu*} \end{bmatrix} := \begin{bmatrix} \alpha(r_1) \otimes D_{11} \otimes \alpha(m_1)' & \cdots & \alpha(r_\nu) \otimes D_{1\nu} \otimes \alpha(m_1)' \\ \vdots & \ddots & \vdots \\ \alpha(r_1) \otimes D_{\nu 1} \otimes \alpha(m_\nu)' & \cdots & \alpha(r_\nu) \otimes D_{\nu\nu} \otimes \alpha(m_\nu)' \end{bmatrix}$$

where $\alpha(p) = \begin{bmatrix} 1 & \cdots & 1 \end{bmatrix} \in R^p$, and where $\otimes$ represents the Kronecker product (the MATLAB command for that is *kron*). The representation (2.23) is called *scalar expanded form* of system (2.1) [3].

Example 2.4 Examples of the Kronecker product are given below:

$$\begin{bmatrix} 1 & 1 & 1 \end{bmatrix} \otimes \begin{bmatrix} 1 & 2 \\ 3 & 4 \end{bmatrix} = \begin{bmatrix} 1 & 2 & 1 & 2 & 1 & 2 \\ 3 & 4 & 3 & 4 & 3 & 4 \end{bmatrix},$$

$$\begin{bmatrix} 1 & 2 \\ 3 & 4 \end{bmatrix} \otimes \begin{bmatrix} 1 \\ 1 \end{bmatrix} = \begin{bmatrix} 1 & 2 \\ 1 & 2 \\ 3 & 4 \\ 3 & 4 \end{bmatrix},$$

$$\begin{bmatrix} 1 & 1 \end{bmatrix} \otimes \begin{bmatrix} 1 & 2 \\ 3 & 4 \end{bmatrix} \otimes \begin{bmatrix} 1 \\ 1 \\ 1 \end{bmatrix} = \begin{bmatrix} 1 & 2 & 1 & 2 \\ 1 & 2 & 1 & 2 \\ 1 & 2 & 1 & 2 \\ 3 & 4 & 3 & 4 \\ 3 & 4 & 3 & 4 \\ 3 & 4 & 3 & 4 \end{bmatrix}$$

Theorem 2.4 [2] *Consider the system given by (2.1). Then,* $\lambda \in \mathrm{sp}(A)$ *is a DFM of the system with respect to the block diagonal information flow* $\mathbf{K}$ *given by* (2.11) if and only if λ *is a transmission zero of all of the following systems:*

(1) (c_i, A, b_i),
$i = 1, 2, \ldots, \nu^*$

(2) $\left(\begin{bmatrix} c_i \\ c_j \end{bmatrix}, A, \begin{bmatrix} b_i & b_j \end{bmatrix}, \begin{bmatrix} 0 & d_{ij} \\ d_{ji} & 0 \end{bmatrix}\right)$,
$i = 1, 2, \ldots, \nu^* - 1;\ j = i+1, i+2, \ldots, \nu^*$

(3) $\left(\begin{bmatrix} c_i \\ c_j \\ c_k \end{bmatrix}, A, \begin{bmatrix} b_i & b_j & b_k \end{bmatrix}, \begin{bmatrix} 0 & d_{ij} & d_{ik} \\ d_{ji} & 0 & d_{jk} \\ d_{ki} & d_{kj} & 0 \end{bmatrix}\right)$,

$i = 1, 2, \ldots, \nu^*-2;\ j = i+1, i+2, \ldots, \nu^*-1;\ k = j+1, j+2, \ldots, \nu^*$

$\vdots$

$$(\nu^*)\quad \left(\begin{bmatrix} c_1 \\ c_2 \\ \vdots \\ c_{\nu^*} \end{bmatrix}, A, \begin{bmatrix} b_1 & b_2 & \cdots & b_{\nu^*} \end{bmatrix}, \begin{bmatrix} 0 & d_{12} & \cdots & d_{1\nu^*} \\ d_{21} & 0 & \cdots & d_{2\nu^*} \\ \vdots & \vdots & \ddots & \vdots \\ d_{\nu^*1} & d_{\nu^*2} & \cdots & 0 \end{bmatrix}\right).$$

Note that the D_{ii} terms do not play any role in the characterization of decentralized fixed modes. This observation is not surprising since the available local information enables one to "remove" the D_{ii} terms from a feedback controller.

Example 2.5 Consider the system (2.1) with $\nu = 2$ and the following system parameters:

$$A = \begin{bmatrix} 1 & 0 & 0 \\ 0 & 2 & 0 \\ 0 & 0 & 3 \end{bmatrix},\ B_1 = \begin{bmatrix} 1 \\ 0 \\ 1 \end{bmatrix},\ B_2 = \begin{bmatrix} 0 & 1 \\ 1 & 0 \\ 0 & -1 \end{bmatrix},$$

$$C_1 = \begin{bmatrix} 0 & 1 & 2 \\ 1 & 2 & 0 \end{bmatrix},\ C_2 = \begin{bmatrix} 1 & 0 & 1 \end{bmatrix},\ D = \begin{bmatrix} 0 & 0 & 0 \\ 0 & 0 & 0 \\ 0 & 0 & 0 \end{bmatrix}. \tag{2.24}$$

We will use both methods to check if this system has any DFM. In this example, we have $\nu = 2$, $m_1 = 1$, $m_2 = 2$, $r_1 = 2$ and $r_2 = 1$. This means that $K_1 \in R^{1\times 2}$ and $K_2 \in R^{2\times 1}$, and so from (2.11) the decentralized information flow $\mathbf{K}$ will have the following form:

$$K = \begin{bmatrix} K_1 & \mathbf{0} \\ \mathbf{0} & K_2 \end{bmatrix} = \begin{bmatrix} \times & \times & 0 \\ 0 & 0 & \times \\ 0 & 0 & \times \end{bmatrix}.$$

The eigenvalues of A are $\mathrm{sp}(A) = \{1,\ 2,\ 3\}$. We want to check if $\lambda = 2$ is a DFM of this system. Using Theorem 2.2, we will have:

(1) Check if $\mathrm{rank}\left(\begin{bmatrix} A - \lambda I \\ C_1 \\ C_2 \end{bmatrix}\right) < 3$

$$\begin{bmatrix} A-\lambda I \\ C_1 \\ C_2 \end{bmatrix} = \begin{bmatrix} -1 & 0 & 0 \\ 0 & 0 & 0 \\ 0 & 0 & 1 \\ 0 & 1 & 2 \\ 1 & 2 & 0 \\ 1 & 0 & 1 \end{bmatrix} \Rightarrow \text{rank}\left(\begin{bmatrix} A-\lambda I \\ C_1 \\ C_2 \end{bmatrix}\right) = 3$$

(2) Check if $\text{rank}\left(\begin{bmatrix} A-\lambda I & B_1 & B_2 \end{bmatrix}\right) < 3$

$$\begin{bmatrix} A-\lambda I & B_1 & B_2 \end{bmatrix} = \begin{bmatrix} -1 & 0 & 0 & 1 & 0 & 1 \\ 0 & 0 & 0 & 0 & 1 & 0 \\ 0 & 0 & 1 & 1 & 0 & -1 \end{bmatrix}$$

$$\Rightarrow \text{rank}\left(\begin{bmatrix} A-\lambda I & B_1 & B_2 \end{bmatrix}\right) = 3$$

(3) Check if $\text{rank}\left(\begin{bmatrix} A-\lambda I & B_1 \\ C_2 & 0 \end{bmatrix}\right) < 3$ or $\text{rank}\left(\begin{bmatrix} A-\lambda I & B_2 \\ C_1 & 0 \end{bmatrix}\right) < 3$

$$\begin{bmatrix} A-\lambda I & B_1 \\ C_2 & 0 \end{bmatrix} = \begin{bmatrix} -1 & 0 & 0 & 1 \\ 0 & 0 & 0 & 0 \\ 0 & 0 & 1 & 1 \\ 1 & 0 & 1 & 0 \end{bmatrix} \Rightarrow \text{rank}\left(\begin{bmatrix} A-\lambda I & B_1 \\ C_2 & 0 \end{bmatrix}\right) = 2 < 3$$

So, the system has a DFM at $\lambda = 2$ (there is no need to check the rank of $\begin{bmatrix} A-\lambda I & B_2 \\ C_1 & 0 \end{bmatrix}$).

We will now use Theorem 2.4 to solve this problem. Here, we have $m_1 = 1$, $m_2 = 2$, $r_1 = 2$, $r_2 = 1$ and $\nu^* = \sum_{i=1}^{\nu} m_i r_i = 2 + 2 = 4$. Therefore,

$$\begin{aligned} \begin{bmatrix} b_1 & b_2 & b_3 & b_4 \end{bmatrix} &= \begin{bmatrix} \alpha(r_1) \otimes B_1 & \alpha(r_2) \otimes B_2 \end{bmatrix} \\ &= \begin{bmatrix} \begin{bmatrix} 1 & 1 \end{bmatrix} \otimes B_1 & [1] \otimes B_2 \end{bmatrix} = \begin{bmatrix} 1 & 1 & 0 & 1 \\ 0 & 0 & 1 & 0 \\ 1 & 1 & 0 & -1 \end{bmatrix} \end{aligned}$$

$$\begin{bmatrix} c_1 \\ c_2 \\ c_3 \\ c_4 \end{bmatrix} = \begin{bmatrix} C_1 \otimes \alpha(m_1)' \\ C_\nu \otimes \alpha(m_2)' \end{bmatrix} = \begin{bmatrix} C_1 \otimes [1] \\ C_\nu \otimes \begin{bmatrix} 1 \\ 1 \end{bmatrix} \end{bmatrix} = \begin{bmatrix} 0 & 1 & 2 \\ 1 & 2 & 0 \\ 1 & 0 & 1 \\ 1 & 0 & 1 \end{bmatrix}$$

(1) Check if $\lambda = 2$ is a transmission zero of (c_i, A, b_i), $i = 1, 2, 3, 4$.

$$\text{rank}\left(\begin{bmatrix} A-\lambda I & b_1 \\ c_1 & 0 \end{bmatrix}\right) = \text{rank}\left(\begin{bmatrix} -1 & 0 & 0 & 1 \\ 0 & 0 & 0 & 0 \\ 0 & 0 & 1 & 1 \\ 0 & 1 & 2 & 0 \end{bmatrix}\right) = 3 < 4,$$

$$\text{rank}\left(\begin{bmatrix} A-\lambda I & b_2 \\ c_2 & 0 \end{bmatrix}\right) = \text{rank}\left(\begin{bmatrix} -1 & 0 & 0 & 1 \\ 0 & 0 & 0 & 0 \\ 0 & 0 & 1 & 1 \\ 1 & 2 & 0 & 0 \end{bmatrix}\right) = 3 < 4,$$

$$\text{rank}\left(\begin{bmatrix} A-\lambda I & b_3 \\ c_3 & 0 \end{bmatrix}\right) = \text{rank}\left(\begin{bmatrix} -1 & 0 & 0 & 0 \\ 0 & 0 & 0 & 1 \\ 0 & 0 & 1 & 0 \\ 1 & 0 & 1 & 0 \end{bmatrix}\right) = 3 < 4,$$

$$\text{rank}\left(\begin{bmatrix} A-\lambda I & b_4 \\ c_4 & 0 \end{bmatrix}\right) = \text{rank}\left(\begin{bmatrix} -1 & 0 & 0 & 1 \\ 0 & 0 & 0 & 0 \\ 0 & 0 & 1 & -1 \\ 1 & 0 & 1 & 0 \end{bmatrix}\right) = 3 < 4.$$

(2) Check if $\lambda = 2$ is a transmission zero of $\left(\begin{bmatrix} c_i \\ c_j \end{bmatrix}, A, \begin{bmatrix} b_i & b_j \end{bmatrix}, \begin{bmatrix} 0 & d_{ij} \\ d_{ji} & 0 \end{bmatrix}\right)$, $i = 1, 2, 3;\ \ j = i+1, \ldots, 4$.

$$\text{rank}\left(\begin{bmatrix} A-\lambda I & b_1 & b_2 \\ c_1 & 0 & 0 \\ c_2 & 0 & 0 \end{bmatrix}\right) = \text{rank}\left(\begin{bmatrix} -1 & 0 & 0 & 1 & 1 \\ 0 & 0 & 0 & 0 & 0 \\ 0 & 0 & 1 & 1 & 1 \\ 0 & 1 & 2 & 0 & 0 \\ 1 & 2 & 0 & 0 & 0 \end{bmatrix}\right) = 4 < 5,$$

$$\text{rank}\left(\begin{bmatrix} A-\lambda I & b_1 & b_3 \\ c_1 & 0 & 0 \\ c_3 & 0 & 0 \end{bmatrix}\right) = \text{rank}\left(\begin{bmatrix} -1 & 0 & 0 & 1 & 0 \\ 0 & 0 & 0 & 0 & 1 \\ 0 & 0 & 1 & 1 & 0 \\ 0 & 1 & 2 & 0 & 0 \\ 1 & 0 & 1 & 0 & 0 \end{bmatrix}\right) = 4 < 5,$$

$$\text{rank}\left(\begin{bmatrix} A-\lambda I & b_1 & b_4 \\ c_1 & 0 & 0 \\ c_4 & 0 & 0 \end{bmatrix}\right) = \text{rank}\left(\begin{bmatrix} -1 & 0 & 0 & 1 & 1 \\ 0 & 0 & 0 & 0 & 0 \\ 0 & 0 & 1 & 1 & -1 \\ 0 & 1 & 2 & 0 & 0 \\ 1 & 0 & 1 & 0 & 0 \end{bmatrix}\right) = 4 < 5,$$

$$\text{rank}\left(\begin{bmatrix} A-\lambda I & b_2 & b_3 \\ c_2 & 0 & 0 \\ c_3 & 0 & 0 \end{bmatrix}\right) = \text{rank}\left(\begin{bmatrix} -1 & 0 & 0 & 1 & 0 \\ 0 & 0 & 0 & 0 & 1 \\ 0 & 0 & 1 & 1 & 0 \\ 1 & 2 & 0 & 0 & 0 \\ 1 & 0 & 1 & 0 & 0 \end{bmatrix}\right) = 4 < 5,$$

$$\text{rank}\left(\begin{bmatrix} A-\lambda I & b_2 & b_4 \\ c_2 & 0 & 0 \\ c_4 & 0 & 0 \end{bmatrix}\right) = \text{rank}\left(\begin{bmatrix} -1 & 0 & 0 & 1 & 1 \\ 0 & 0 & 0 & 0 & 0 \\ 0 & 0 & 1 & 1 & -1 \\ 1 & 2 & 0 & 0 & 0 \\ 1 & 0 & 1 & 0 & 0 \end{bmatrix}\right) = 4 < 5,$$

$$\text{rank}\left(\begin{bmatrix} A-\lambda I & b_3 & b_4 \\ c_3 & 0 & 0 \\ c_4 & 0 & 0 \end{bmatrix}\right) = \text{rank}\left(\begin{bmatrix} -1 & 0 & 0 & 0 & 1 \\ 0 & 0 & 0 & 1 & 0 \\ 0 & 0 & 1 & 0 & -1 \\ 1 & 0 & 1 & 0 & 0 \\ 1 & 0 & 1 & 0 & 0 \end{bmatrix}\right) = 4 < 5$$

(3) Check if $\lambda = 2$ is a transmission zero of $\left(\begin{bmatrix} c_i \\ c_j \\ c_k \end{bmatrix}, A, \begin{bmatrix} b_i & b_j & b_k \end{bmatrix}, \begin{bmatrix} 0 & d_{ij} & d_{ik} \\ d_{ji} & 0 & d_{jk} \\ d_{ki} & d_{kj} & 0 \end{bmatrix}\right)$,
$i = 1, 2;\ j = i+1, \ldots, 3;\ k = j+1, \ldots, 4.$

$$\text{rank}\left(\begin{bmatrix} A-\lambda I & b_1 & b_2 & b_3 \\ c_1 & 0 & 0 & 0 \\ c_2 & 0 & 0 & 0 \\ c_3 & 0 & 0 & 0 \end{bmatrix}\right) = \text{rank}\left(\begin{bmatrix} -1 & 0 & 0 & 1 & 1 & 0 \\ 0 & 0 & 0 & 0 & 0 & 1 \\ 0 & 0 & 1 & 1 & 1 & 0 \\ 0 & 1 & 2 & 0 & 0 & 0 \\ 1 & 2 & 0 & 0 & 0 & 0 \\ 1 & 0 & 1 & 0 & 0 & 0 \end{bmatrix}\right) = 5 < 6,$$

$$\text{rank}\left(\begin{bmatrix} A-\lambda I & b_1 & b_2 & b_4 \\ c_1 & 0 & 0 & 0 \\ c_2 & 0 & 0 & 0 \\ c_4 & 0 & 0 & 0 \end{bmatrix}\right) = \text{rank}\left(\begin{bmatrix} -1 & 0 & 0 & 1 & 1 & 1 \\ 0 & 0 & 0 & 0 & 0 & 0 \\ 0 & 0 & 1 & 1 & 1 & -1 \\ 0 & 1 & 2 & 0 & 0 & 0 \\ 1 & 2 & 0 & 0 & 0 & 0 \\ 1 & 0 & 1 & 0 & 0 & 0 \end{bmatrix}\right) = 5 < 6$$

$$\text{rank}\left(\begin{bmatrix} A-\lambda I & b_2 & b_3 & b_4 \\ c_2 & 0 & 0 & 0 \\ c_3 & 0 & 0 & 0 \\ c_4 & 0 & 0 & 0 \end{bmatrix}\right) = \text{rank}\left(\begin{bmatrix} -1 & 0 & 0 & 1 & 0 & 1 \\ 0 & 0 & 0 & 0 & 1 & 0 \\ 0 & 0 & 1 & 1 & 0 & -1 \\ 1 & 2 & 0 & 0 & 0 & 0 \\ 1 & 0 & 1 & 0 & 0 & 0 \\ 1 & 0 & 1 & 0 & 0 & 0 \end{bmatrix}\right) = 5 < 6$$

(4) Check if $\lambda = 2$ is a transmission zero of $\left(\begin{bmatrix} c_1 \\ c_2 \\ c_3 \\ c_4 \end{bmatrix}, A, \begin{bmatrix} b_1 & b_2 & b_3 & b_4 \end{bmatrix}, \begin{bmatrix} 0 & d_{12} & d_{13} & d_{14} \\ d_{21} & 0 & d_{23} & d_{24} \\ d_{31} & d_{32} & 0 & d_{34} \\ d_{41} & d_{42} & d_{43} & 0 \end{bmatrix}\right)$

$$\text{rank}\left(\begin{bmatrix} A-\lambda I & b_1 & b_2 & b_3 & b_4 \\ c_1 & 0 & 0 & 0 & 0 \\ c_2 & 0 & 0 & 0 & 0 \\ c_3 & 0 & 0 & 0 & 0 \\ c_4 & 0 & 0 & 0 & 0 \end{bmatrix}\right) = \text{rank}\left(\begin{bmatrix} -1 & 0 & 0 & 1 & 1 & 0 & 1 \\ 0 & 0 & 0 & 0 & 0 & 1 & 0 \\ 0 & 0 & 1 & 1 & 1 & 0 & -1 \\ 0 & 1 & 2 & 0 & 0 & 0 & 0 \\ 1 & 2 & 0 & 0 & 0 & 0 & 0 \\ 1 & 0 & 1 & 0 & 0 & 0 & 0 \\ 1 & 0 & 1 & 0 & 0 & 0 & 0 \end{bmatrix}\right) = 6 < 7$$

Since $\lambda = 2$ is a transmission zero of all systems given in (1)–(4), it is a DFM of (2.1).

2.3.3 Relaxing Generic Gain Condition

The following algorithm gives a procedure to determine the DFM of the system (2.1), which does not require the choice of a generic gain matrix to be made.

An Algorithm to determine the DFM of (2.1) *using a finite number of gain matrices* [2]: The DFM can be determined by applying only a *finite* number of arbitrary output gain matrices $K(i)$, $i = 1, 2, \ldots, p$ (where $K(i)$ have a specific structure) to the system and then determining the system's eigenvalues. Assume with no loss of generality that (2.1) is written in the form of (2.23), and let:

$$\overline{C} := \begin{bmatrix} c_1 \\ \vdots \\ c_{\nu^*} \end{bmatrix}, \quad \overline{B} := \begin{bmatrix} b_1 & \cdots & b_{\nu^*} \end{bmatrix}, \quad \overline{D} := \begin{bmatrix} d_{11} & \cdots & d_{1\nu^*} \\ \vdots & \ddots & \vdots \\ d_{\nu^*1} & \cdots & d_{\nu^*\nu^*} \end{bmatrix}$$

Let also $p := \sum_{i=1}^{\nu^*} \frac{\nu^*!}{i!(\nu^*-i)!}$, and define the following diagonal gain matrices $K(i)$, $i = 1, 2, \ldots, p$ for (2.23). $K(1), \ldots, K(p)$ are generated by constructing a total of p matrices consisting of all matrices which have a single nonzero element, two nonzero elements, ..., ν^* nonzero elements; i.e.,

$$\begin{aligned}
&K(1) := \text{diag}\big(\big[\times\ 0\ 0 \cdots 0\ 0\ 0\big]\big) \\
&\vdots \\
&K(\nu^*) := \text{diag}\big(\big[0\ 0\ 0 \cdots 0\ 0\ \times\big]\big) \\
&K(\nu^*+1) := \text{diag}\big(\big[\times\ \times\ 0 \cdots 0\ 0\ 0\big]\big) \\
&\vdots \\
&K(\nu^*(\nu^*+1)/2) := \text{diag}\big(\big[0\ 0\ 0 \cdots 0\ \times\ \times\big]\big) \\
&\vdots \\
&K(p) := \text{diag}\big(\big[\times\ \times\ \times \cdots \times\ \times\ \times\big]\big)
\end{aligned}$$

where diag$([\, w_1 \cdots w_q \,])$ represents a diagonal matrix whose (i, i) element is equal to w_i, and $\times$ are arbitrary nonzero scalars chosen so that $\det[(I - \overline{D}K(i))] \neq 0$, $i = 1, 2, \ldots, p$.

Let $A_c(i)$, $i = 1, 2, \ldots, p$ be defined as follows:

$$A_c(i) := A + \overline{B}K(i)(I - \overline{D}K(i))^{-1}\overline{C}, \quad i = 1, 2, \ldots, p.$$

The following result is obtained.

Proposition 2.1 [2] *Given the system (2.23), the set of DFMs with respect to the diagonal information flow* **K** *is equal to:*

$$\mathrm{sp}(A) \bigcap_{i=1}^{p} \mathrm{sp}(A_c(i)).$$

2.4 Decentralized Control Synthesis

2.4.1 Decentralized Pole Assignment

Consider the system (2.1), and let $\lambda_i \in \mathrm{sp}(A)$, $i = 1, 2, \ldots, n$. Let also C_g^- be a prescribed symmetric (with respect to the real axis) open region of the complex plane.

Theorem 2.5 [2] *There exists a decentralized LTI controller for (2.1) such that all poles of the closed-loop system are contained in C_g^- if and only if the DFMs of (2.1) are contained in C_g^-.*

A decentralized pole assignment method is given in the sequel, using centralized synthesis. Let

$$B^k := [\, B_1 \cdots B_k \,],\; C^k := \begin{bmatrix} C_1 \\ \vdots \\ C_k \end{bmatrix},\; D^k := \begin{bmatrix} D_{11} & \cdots & D_{1k} \\ \vdots & \ddots & \vdots \\ D_{k1} & \cdots & D_{kk} \end{bmatrix}.$$

Given (2.1), consider the following decentralized controller P_i:

$$P_i : \begin{cases} \dot{z}_i(t) = S_i z_i(t) + R_i y_i(t) \\ u_i(t) = Q_i z_i(t) + K_i y_i(t) \end{cases}, \quad i = 1, 2, \ldots, \nu$$

where $z_i(t) \in R^{\eta_i}$ is the state vector of the ith feedback controller, and define:

$$S^k := \begin{bmatrix} S_1 & & \mathbf{0} \\ & \ddots & \\ \mathbf{0} & & S_k \end{bmatrix}, \quad R^k := \begin{bmatrix} R_1 & & \mathbf{0} \\ & \ddots & \\ \mathbf{0} & & R_k \end{bmatrix}$$
$$Q^k := \begin{bmatrix} Q_1 & & \mathbf{0} \\ & \ddots & \\ \mathbf{0} & & Q_k \end{bmatrix}, \quad K^k := \begin{bmatrix} K_1 & & \mathbf{0} \\ & \ddots & \\ \mathbf{0} & & K_k \end{bmatrix}, \quad k = 1, 2, \ldots, \nu.$$

Assume $\det(I - D^k K^k) \neq 0$ and define the closed-loop system matrix A_k, $k = 1, 2, \ldots, \nu$ obtained by applying the controller P_i, $i = 1, 2, \ldots, k$ to (2.1) as follows:

$$A_k := \begin{bmatrix} A + B^k(I - K^k D^k)^{-1} K^k C^k & B^k(I - K^k D^k)^{-1} Q^k \\ R^k[I + D^k(I - K^k D^k)^{-1} K^k] C^k & S^k + R^k D^k (I - K^k D^k)^{-1} Q^k \end{bmatrix}.$$

Then, the resultant system for local control agent $k + 1$ with input u_{k+1} and output y_{k+1}, $k = 1, 2, \ldots, \nu - 1$, is described by:

$$\begin{bmatrix} \dot{x}(t) \\ \dot{z}_1(t) \\ \vdots \\ \dot{z}_k(t) \end{bmatrix} = A_k \begin{bmatrix} x(t) \\ z_1(t) \\ \vdots \\ z_k(t) \end{bmatrix} + \overline{B}_k u_{k+1}(t)$$

$$y_{k+1}(t) = \overline{C}_k \begin{bmatrix} x(t) \\ z_1(t) \\ \vdots \\ z_k(t) \end{bmatrix} + \overline{D}_k u_{k+1}(t)$$

where $\overline{B}_k$, $\overline{C}_k$ and $\overline{D}_k$ are given by:

$$\overline{B}_k := \begin{bmatrix} B_{k+1} + B^k(I - K^k D^k)^{-1} K^k \begin{bmatrix} D_{1,k+1} \\ \vdots \\ D_{k,k+1} \end{bmatrix} \\ R^k[I + D^k(I - K^k D^k)^{-1} K^k] \begin{bmatrix} D_{1,k+1} \\ \vdots \\ D_{k,k+1} \end{bmatrix} \end{bmatrix}$$

$$\overline{C}_k := \Big[\, C_{k+1} + \big[\, D_{k+1,1} \cdots D_{k+1,k} \,\big](I - K^k D^k)^{-1} K^k C^k \;\; \big[\, D_{k+1,1} \cdots D_{k+1,k} \,\big](I - K^k D^k)^{-1} Q^k \,\Big]$$

$$\overline{D}_k := D_{k+1,k+1} + \big[\, D_{k+1,1} \cdots D_{k+1,k} \,\big](I - K^k D^k)^{-1} K^k \begin{bmatrix} D_{1,k+1} \\ \vdots \\ D_{k,k+1} \end{bmatrix}$$

The following algorithm can be used to construct a decentralized LTI controller for (2.1) which satisfies Theorem 2.4 [2]:

(1) Using centralized synthesis, design the stabilizing compensator P_1 so that the closed-loop system matrix A_1 has all of its eigenvalues, except for those eigenvalues which correspond to the CFMs of (C_1, A, B_1), contained in C_g^-. *Note that the CFMs of* (C_1, A, B_1) *are, in fact, the DFMs of* (C^ν, A, B^ν, D^ν) *with respect to* $\text{diag}\big(\big[\times\ 0\ 0 \cdots 0\ 0\ 0\big]\big)$.

(2) Using centralized synthesis, design the stabilizing compensator P_2 so that the closed-loop system matrix A_2 has all of its eigenvalues, except for those eigenvalues which correspond to the DFMs of (C^2, A, B^2, D^2), contained in C_g^-. This is possible to do for almost all K_1, Q_1. *Note that the DFMs of* (C^2, A, B^2, D^2) *are, in fact, the DFMs of* (C^ν, A, B^ν, D^ν) *with respect to* $\text{diag}\big(\big[\times\ \times\ 0 \cdots 0\ 0\ 0\big]\big)$.

$\vdots$

(k) Using centralized synthesis, design the stabilizing compensator P_k so that the closed-loop system matrix A_k has all of its eigenvalues, except for those eigenvalues which correspond to the DFMs of (C^k, A, B^k, D^k), contained in C_g^-. This is possible to do for almost all K_i, Q_i, $i = 1, 2, \ldots, k-1$.

$\vdots$

(ν) Using centralized synthesis, design the stabilizing compensator P_ν so that the closed-loop system matrix A_ν has all of its eigenvalues, except for those eigenvalues which correspond to the DFMs of (C^ν, A, B^ν, D^ν), contained in C_g^-. This is possible to do for almost all K_i, Q_i, $i = 1, 2, \ldots, \nu - 1$.

In this procedure, it may turn out that the minimal realization of the centralized system at step i, $1, 2, \ldots, q$, is identically zero or has a minimal realization of order 0; if this is the case, it is essential that the stabilizing compensator $u_i(t) = K_i y_i(t)$, $i = 1, 2, \ldots, q$ still be applied, where K_i are arbitrary nonzero gain matrices, in order for the algorithm to proceed. It is to be noted that the order of the stabilizing compensator P_k required, in general, increases as k increases.

2.5 Summary

In this chapter, the stability of decentralized control systems is investigated, which is shown to be dependent on the notion of decentralized fixed modes (DFMs). Different techniques are provided to characterize the DFMs of a general proper finite-dimensional LTI system. The first technique is an efficient numerical method which can be easily utilized on MATLAB. It basically uses the random number generator to check if any of the eigenvalues of the system remains fixed with respect to a generic constant gain matrix of the desired decentralized structure. The other techniques are analytical and need to verify the rank of a set of matrices obtained from the system matrices. One of the techniques requires checking the transmission zeros of a set of systems. The technique is initially given for a decentralized control system with SISO subsystems and is then extended to the systems with MIMO subsystems. The main characteristic of this method is that it requires the transmission zero of every (virtual) system in a set to be checked, which implies that the ranks of all of the corresponding matrices need to be verified. The other method, on the other hand, requires the ranks of a different set of matrices to be checked. The main difference of this method with the former one is that if any of the rank conditions does not hold, the corresponding eigenvalue is a DFM and there is no need to check the remaining matrices. Furthermore, a numerical method is provided which unlike the one described earlier does not require the choice of generic gain matrices. Finally, a procedure is introduced to find an LTI decentralized controller of proper structure to locate the poles of the closed-loop system in desired locations in the complex plane. The procedure uses centralized design techniques to find the local controllers.

2.6 Problems

1. Consider the temperature control system whose parameters are given by build20.mat. Using random gain matrices, find the DFMs with respect to:

   (a) $K_1 = \begin{bmatrix} \times & 0 & 0 & 0 \\ 0 & \times & 0 & 0 \\ 0 & 0 & \times & 0 \\ 0 & 0 & 0 & \times \end{bmatrix}$

   (b) $K_2 = \begin{bmatrix} 0 & 0 & 0 & \times \\ 0 & 0 & \times & 0 \\ 0 & \times & 0 & 0 \\ \times & 0 & 0 & 0 \end{bmatrix}$

2. Consider a decentralized system of the form (2.2), where

$$A = \begin{bmatrix} 1 & 0 & 0 \\ 0 & 1 & 0 \\ 0 & 0 & 2 \end{bmatrix}, B_1 = \begin{bmatrix} 1 \\ 0 \\ 1 \end{bmatrix}, B_2 = \begin{bmatrix} 0 \\ 1 \\ 1 \end{bmatrix}, C_1 = \begin{bmatrix} 0 & 0 & 1 \end{bmatrix}, C_2 = \begin{bmatrix} 1 & 0 & 1 \\ 0 & 1 & 1 \end{bmatrix},$$

$$D_{11} = D_{12} = 0, D_{21} = D_{22} = \begin{bmatrix} 0 \\ 0 \end{bmatrix}.$$

Use Theorems 2.2 and 2.3 (with MIMO subsystems) to find the DFMs of this system, if any.

3. Consider an LTI system:

$$\dot{x}(t) = Ax(t) + Bu(t)$$
$$y(t) = Cx(t)$$

with $x(t) \in R^n$, $u(t) \in R^3$ and $y(t) \in R^3$. Show that there exist gain matrices $K(i) \in R^{3\times 3}$, $i = 1, 2, \ldots, N$, such that only N tests on the eigenvalues of the closed-loop matrix $A + BK(i)C$, $i = 1, 2, \ldots, N$, are required in order to determine if the system (C, A, B) has any DFMs with respect to $K_1 = \begin{bmatrix} \times & 0 & 0 \\ 0 & \times & 0 \\ 0 & 0 & \times \end{bmatrix}$. Find the minimum value of N which allows this to be achieved and find $K(i)$, $i = 1, 2, \ldots, N$.

4. Repeat Problem 3 for the case when "decentralized fixed modes" is replaced by "centralized fixed modes".

5. Consider the following system:

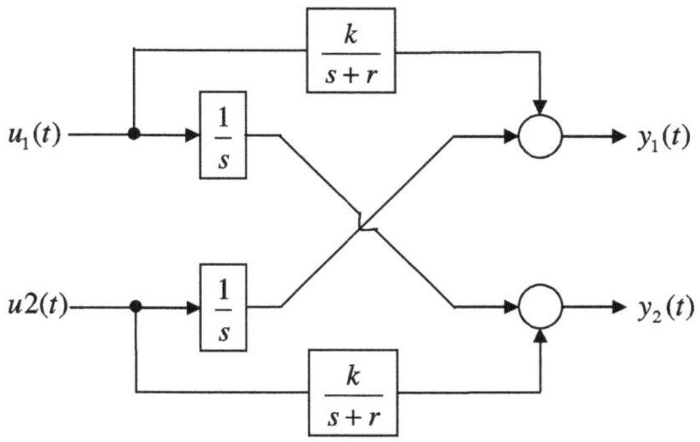

Determine if the system can be stabilized using a decentralized LTI controller with control agents:

(a) $(u_1(t), y_1(t))$ and $(u_2(t), y_2(t))$, i.e., $K = \begin{bmatrix} \times & 0 \\ 0 & \times \end{bmatrix}$.

(b) $(u_1(t), y_2(t))$ and $(u_2(t), y_1(t))$, i.e., $K = \begin{bmatrix} 0 & \times \\ \times & 0 \end{bmatrix}$.

6. Consider the following system:

$$\dot{x}(t) = Ax(t) + \begin{bmatrix} b_1 & b_2 & b_3 \end{bmatrix} \begin{bmatrix} u_1(t) \\ u_2(t) \\ u_3(t) \end{bmatrix}$$

$$\begin{bmatrix} y_1(t) \\ y_2(t) \\ y_3(t) \end{bmatrix} = \begin{bmatrix} c_1 \\ c_2 \\ c_3 \end{bmatrix} x(t) + \begin{bmatrix} d_{11} & d_{12} & d_{13} \\ d_{21} & d_{22} & d_{23} \\ d_{31} & d_{32} & d_{33} \end{bmatrix} \begin{bmatrix} u_1(t) \\ u_2(t) \\ u_3(t) \end{bmatrix}$$

where $A \in R^{n \times n}$, b_1, b_2, b_3 are column vectors and c_1, c_2, c_3 are row vectors. Find a set of conditions similar to the ones given in Theorem 2.2 to check if the system has any DFM with respect to the following information flow structure:

$$K = \begin{bmatrix} 0 & 0 & \times \\ \times & \times & 0 \\ \times & \times & 0 \end{bmatrix}.$$

7. Consider the following system:

$$\dot{x}(t) = Ax(t) + \begin{bmatrix} b_1 & b_2 \end{bmatrix} \begin{bmatrix} u_1(t) \\ u_2(t) \end{bmatrix}$$
$$y(t) = cx(t)$$

where b_1 , $b_2 \in R^{2 \times 1}$, and $c \in R^{1 \times 2}$. Prove that $\lambda \in \text{sp}(A)$ is an unobservable *AND* uncontrollable mode of this system if and only if it is a transmission zero of a set of systems. Show all of these systems.

8. Consider a set of four micro satellites S_1, S_2, S_3, S_4, and assume that the dynamic equations of the satellite S_i, $i \in \{1, 2, 3, 4\}$, are given by:

$$\dot{x}_i(t) = A_i x_i(t) + b_i u_i(t)$$

Assume also that the position of the satellite S_i in a given frame is described by:

$$p_i(t) = h_i x_i(t)$$

Note that in general $p_i(t) \in R^3$, but for simplicity, suppose that $p_i(t) \in R$ (i.e. all satellites are moving on a straight line).

(a) Write the equations of the overall system in the following form:

$$\dot{x}(t) = Ax(t) + Bu(t)$$
$$y(t) = Cx(t)$$

Here, $y_i(t) \in R$ is the relative position of the satellite S_i with respect to S_{i+1}, $i \in \{1, 2, 3, 4\}$, where S_5 is defined to be the same as S_1.

(b) Assume that it is desired to apply a decentralized controller to this system such that the local controller K_i associated with the satellite S_i uses the information of the satellites S_i, S_{i+1} and S_{i-1} in order to compute the corresponding control command, where S_0 and S_5 are defined to be the same as S_4 and S_1, respectively. Find the information flow structure for the corresponding decentralized controller.

(c) Find a set of new equations

$$\dot{x}(t) = Ax(t) + \overline{B}\overline{u}(t)$$
$$\overline{y}(t) = \overline{C}x(t)$$

such that the decentralized controller defined in part (b) has a diagonal information flow structure in the new set-up.

9. Consider the following system:

$$\dot{x}(t) = Ax(t) + \begin{bmatrix} b_1 & b_2 \end{bmatrix} \begin{bmatrix} u_1(t) \\ u_2(t) \end{bmatrix}$$
$$\begin{bmatrix} y_1(t) \\ y_2(t) \end{bmatrix} = \begin{bmatrix} c_1 \\ c_2 \end{bmatrix} x(t)$$

where $b_1, b_2 \in R^{3\times 1}$, and $c_1, c_2 \in R^{1\times 3}$.

(a) Prove that $\lambda \in \mathrm{sp}(A)$ is an unobservable mode of this system if and only if it is a transmission zero of a set of systems. Show only two of these systems: one with the minimum number of inputs and outputs and one with the maximum number of inputs and outputs.

(b) It can be shown that some of the systems in part (a) will be redundant. In other words, any $\lambda \in \mathrm{sp}(A)$ will be a transmission zero of those systems, and thus, it is not required to check the transmission zeros of all of those systems. Find the minimum number of systems whose transmission zeros are required to be checked in this case (i.e. remove any redundant system from the set of the systems whose transmission zeros are required to be checked, and specify the systems in the resultant minimal set).

References

1. S.H. Wang, E.J. Davison, On the stabilization of decentralized control systems. IEEE Trans. Autom. Control **AC-18**(5), 473–478 (1973)
2. E.J. Davison, T.N. Chang, Decentralized stabilization and pole assignment for general proper systems. IEEE Trans. Autom. Control **AC-35**(6), 652–664 (1990)
3. A.F. Vaz, E.J. Davison, On the quantitative characterization of approximate decentralized fixed modes using transmission zeros. Mathematics of Control Signal, and Systems **2**, 287–302 (1989)

Chapter 3
Robust Servomechanism Problem

3.1 Robust Servomechanism Problem Statement

Consider an LTI system with ν local control stations, which is represented by:

$$\begin{aligned}\dot{x}(t) &= Ax(t) + \sum_{i=1}^{\nu} B_i u_i(t) + E\omega(t) \\ y_i(t) &= C_i x(t) + D_i u_i(t) + F_i \omega(t), \quad i = 1, \ldots, \nu \\ y_i^{\mathrm{m}}(t) &= C_i^{\mathrm{m}} x(t) + D_i^{\mathrm{m}} u_i(t) + F_i^{\mathrm{m}} \omega(t), \quad i = 1, \ldots, \nu\end{aligned} \tag{3.1}$$

where $x(t) \in R^n$ is the state vector, $u_i(t) \in R^{m_i}$ and $y_i(t) \in R^{r_i}$ are the input and output to be regulated, respectively, of the ith local control station ($i = 1, \ldots, \nu$); $y_i^{\mathrm{m}}(t) \in R^{r_i^{\mathrm{m}}}$ ($i = 1, \ldots, \nu$) is an output which is available for measurement at the ith control station, and $\omega(t) \in R^{\Omega}$ is the disturbance. The matrices $A_i \in R^{n \times m}$, $B_i \in R^{n \times m_i}$, $C_i \in R^{r_i \times n}$, $D_i \in R^{r_i \times m_j}$, $F_i \in R^{r_i \times \Omega}$, $C_i^{\mathrm{m}} \in R^{r_i^{\mathrm{m}} \times n}$, $D_i^{\mathrm{m}} \in R^{r_i^{\mathrm{m}} \times m_j}$, and $F_i^{\mathrm{m}} \in R^{r_i^{\mathrm{m}} \times \Omega}$ ($i = 1, \ldots, \nu$) are constant matrices. Let the reference input for the ith local control station be denoted by $y_i^{\mathrm{ref}}(t) \in R^{r_i}$ ($i = 1, \ldots, \nu$). Then, the error in the ith local control station which is the difference between the output $y_i(t)$ and the specified reference input $y_i^{\mathrm{ref}}(t)$ ($i = 1, \ldots, \nu$) is given by:

$$e_i(t) = y_i(t) - y_i^{\mathrm{ref}}(t), \quad i = 1, \ldots, \nu \tag{3.2}$$

Define:

$$y_{\mathrm{ref}}(t) := \begin{bmatrix} y_1^{\mathrm{ref}}(t) \\ \vdots \\ y_\nu^{\mathrm{ref}}(t) \end{bmatrix}, \quad u(t) := \begin{bmatrix} u_1(t) \\ \vdots \\ u_\nu(t) \end{bmatrix}, \quad e(t) := \begin{bmatrix} e_1(t) \\ \vdots \\ e_\nu(t) \end{bmatrix}, \quad F := \begin{bmatrix} F_1 \\ \vdots \\ F_\nu \end{bmatrix}. \tag{3.3}$$

E. J. Davison et al., *Decentralized Control of Large-Scale Systems*,
https://doi.org/10.1007/978-1-4419-6014-6_3

The disturbance vector $\omega(t)$ is assumed to satisfy the following equation:

$$\begin{aligned}\dot{z}_1(t) &= \overline{A}_1 z_1(t)\\ \omega(t) &= \overline{C}_1 z_1(t)\end{aligned} \tag{3.4}$$

where $z_1(t) \in R^{\overline{n}_1}$ and where the pair $(\overline{C}_1, \overline{A}_1)$ is observable and $z_1(0)$ may or may not be known. This implies that the disturbance signal is a linear combination of exponential functions of time (which include constant signals and sinusoids). The reference input $y_{\text{ref}}(t)$ is assumed to satisfy the following equation:

$$\begin{aligned}\dot{z}_2(t) &= \overline{A}_2 z_2(t)\\ \rho(t) &= \overline{C}_2 z_2(t)\\ y_{\text{ref}}(t) &= G\rho(t)\end{aligned} \tag{3.5}$$

where $z_2(t) \in R^{\overline{n}_2}$ and where the pair $(\overline{C}_2, \overline{A}_2)$ is observable and $z_2(0)$ is known. Similar to the previous case, this implies that the reference signal is a linear combination of exponential functions of time (which include constant signals and sinusoids). Let the minimal polynomials of $\overline{A}_1, \overline{A}_2$ be denoted by $\Lambda_1(s), \Lambda_2(s)$, respectively, and let the zeros of the least common multiple of $\Lambda_1(s), \Lambda_2(s)$ (multiplicities included) be given by:

$$(\lambda_1, \lambda_2, \dots, \lambda_q) \tag{3.6}$$

Let the coefficients α_j, $j = 1, \dots, q$, be given by the coefficients of the following polynomial:

$$\lambda^q + \alpha_q \lambda^{q-1} + \cdots + \alpha_2 \lambda + \alpha_1 = \prod_{i=1}^{q} (\lambda - \lambda_i) \tag{3.7}$$

and define the following $q \times q$ companion matrix:

$$P = \begin{bmatrix} 0 & 1 & 0 & \cdots & 0\\ 0 & 0 & 1 & \cdots & 0\\ \vdots & \vdots & \vdots & & \vdots\\ -\alpha_1 & -\alpha_2 & -\alpha_3 & \cdots & -\alpha_q \end{bmatrix} \tag{3.8}$$

(note that the eigenvalues of P are equal to $\lambda_1, \lambda_2, \dots, \lambda_q$). The following matrices and notations will be used throughout this chapter:

$$B := \begin{bmatrix} B_1 & \cdots & B_\nu \end{bmatrix}, C := \begin{bmatrix} C_1 \\ \vdots \\ C_\nu \end{bmatrix}, D := \begin{bmatrix} D_1 & & \mathbf{0} \\ & \ddots & \\ \mathbf{0} & & D_\nu \end{bmatrix},$$

$$C_{\mathrm{m}} := \begin{bmatrix} C_1^{\mathrm{m}} \\ \vdots \\ C_\nu^{\mathrm{m}} \end{bmatrix}, D_{\mathrm{m}} := \begin{bmatrix} D_1^{\mathrm{m}} & & \mathbf{0} \\ & \ddots & \\ \mathbf{0} & & D_\nu^{\mathrm{m}} \end{bmatrix},$$

$$m := \sum_{i=1}^{\nu} m_i, r := \sum_{i=1}^{\nu} r_i, r^{\mathrm{m}} := \sum_{i=1}^{\nu} r_i^{\mathrm{m}}. \tag{3.9}$$

Assume, with no loss of generality, that $\mathrm{rank}(C_i) = r_i, \mathrm{rank}(B_i) = m_i, i = 1, \ldots, \nu$.

Definition 3.1 [1] Given $\varepsilon > 0$, define Ω_ε as the set of all real matrices $M = \{m_{ij}\}$, whose elements have a magnitude less than ε, i.e.:

$$\Omega_\varepsilon = \left\{ m_{ij} \middle| \ \left|m_{ij}\right| < \varepsilon \right\}$$

Robust centralized (decentralized) servomechanism problem [1]: Given the system (3.1), assume there exists a controller so that:

(1) The resultant controlled system is stable, i.e. $x(t) \to 0$ as $t \to \infty$, $\forall x(0) \in R^n$ with $z_1(0) = 0$, $z_2(0) = 0$.
(2) Asymptotic regulation takes place, i.e. $e(t) \to 0$ as $t \to \infty$, $\forall x(0) \in R^n$, $\forall z_1(0) \in R^{\overline{n}_1}$, $\forall z_2(0) \in R^{\overline{n}_2}$.

and that (1), (2) occur for all initial conditions of the centralized (decentralized) controller state variables. Let the plant parameters A, B, C now be perturbed, i.e. $A \to A + \delta A$, $B \to B + \delta B$, $C \to C + \delta C$ and let $\varepsilon > 0$ be chosen so that the resultant closed-loop system remains stable $\forall \delta A \in \Omega_\varepsilon$, $\forall \delta B \in \Omega_\varepsilon$, $\forall \delta C \in \Omega_\varepsilon$; then if asymptotic regulation still takes place $\forall \delta A \in \Omega_\varepsilon$, $\forall \delta B \in \Omega_\varepsilon$, $\forall \delta C \in \Omega_\varepsilon$, the controller is said to be a robust centralized (decentralized) controller, and the problem of finding a controller to achieve these requirements is called the *robust centralized (decentralized) servomechanism problem.*

Assume, with no loss of generality, that

$$\mathrm{rank}\left(\begin{bmatrix} E \\ F \end{bmatrix}\right) = \mathrm{rank}(\overline{C}_1) = \Omega$$

$$\mathrm{rank}(G) = \mathrm{rank}(\overline{C}_2) = \dim(\rho)$$

In addition, it is assumed for non-triviality that $\max(\Omega, \dim(\rho)) \geq 1$ (which means that the effect of disturbance in the system is not identical to zero), and that all eigenvalues of $\overline{A}_1$ and $\overline{A}_2$ are located in the closed right-half complex plane (which implies that the reference and disturbance signals do not go to zero as time increases. Note that the output of a stable system with decaying input signals goes to zero as time increases).

3.2 Robust Centralized Servomechanism Problem

Lemma 3.1 [1] *A necessary and sufficient condition for the existence of a robust centralized controller for* (3.1) *such that the error* $e(t)$ *given by* (3.2) *approaches zero as* $t \to \infty$ *for all measurable or unmeasurable disturbances described by* (3.4), *for all* $y_{\text{ref}}(t)$ *described by* (3.5), *and such that the closed-loop system is stable is that the following conditions all hold:*

(1) *The triple* (C_{m}, A, B) *is stabilizable and detectable.*

(2) $\text{rank}\left(\begin{bmatrix} A - \lambda_j I & B \\ C & D \end{bmatrix}\right) = n + r, \quad j = 1, \ldots, q$

(3) $y_i^{\text{m}}(t), \quad i = 1, \ldots, \nu$, *must contain the output* $y_i(t), \quad i = 1, \ldots, \nu$, *respectively, i.e.* $y_i(t), \quad i = 1, \ldots, \nu$, *can be measured.*

Control structure [1]: Assume that conditions (1)–(3) given in Lemma 3.1 are satisfied; then any robust controller which regulates (3.1) must have the following structure:

$$u(t) = \hat{K}\hat{x}(t) + Qz(t)$$

where $z(t) \in R^{r \cdot q}$ is the output of the general servocompensator given by:

$$\dot{z}(t) = S^* z(t) + R^* e(t)$$

with:

$$S^* = \underbrace{\begin{bmatrix} P & & \mathbf{0} \\ & \ddots & \\ \mathbf{0} & & P \end{bmatrix}}_{r \text{ blocks}},$$

and $R^* \in R^{r \cdot q \times r}$ is a real matrix of rank r with the property that (S^*, R^*) is controllable; $\hat{x}(t)$ is the output of a stabilizing compensator with inputs $y_{\text{m}}(t)$, $u(t)$,

$z(t)$, where Q, $\hat{K}$ and the parameters of the stabilizing compensator are found using standard methods, e.g. observer theory, to stabilize the following stabilizable and detectable system:

$$\begin{bmatrix} \dot{x}(t) \\ \dot{z}(t) \end{bmatrix} = \begin{bmatrix} A & 0 \\ R^*C & S^* \end{bmatrix} \begin{bmatrix} x(t) \\ z(t) \end{bmatrix} + \begin{bmatrix} B \\ R^*D \end{bmatrix} u(t)$$

$$\begin{bmatrix} y_{\mathrm{m}}(t) \\ z(t) \end{bmatrix} = \begin{bmatrix} C_{\mathrm{m}} & 0 \\ 0 & I \end{bmatrix} \begin{bmatrix} x(t) \\ z(t) \end{bmatrix} + \begin{bmatrix} D_{\mathrm{m}} \\ 0 \end{bmatrix} u(t)$$

3.3 Robust Decentralized Servomechanism Problem

It is desired now to find the existence conditions for the solution of robust decentralized servomechanism problem. Given system (3.1), define:

$$\hat{C}_1 := \begin{bmatrix} C_1^{\mathrm{m}} & 0 & 0 \cdots 0 \\ 0 & I_{r_1} & 0 \cdots 0 \end{bmatrix},$$

$$\hat{C}_2 := \begin{bmatrix} C_2^{\mathrm{m}} & 0 & 0 & \cdots 0 \\ 0 & 0 & I_{r_2} & \cdots 0 \end{bmatrix},$$

$$\vdots$$

$$\hat{C}_\nu := \begin{bmatrix} C_\nu^{\mathrm{m}} & 0 & 0 \cdots & 0 \\ 0 & 0 & 0 \cdots & I_{r_\nu} \end{bmatrix},$$

and let $\hat{C}_{\mathrm{m}} \in R^{(r_{\mathrm{m}}+r)\times(n+r)}$ be given by:

$$\hat{C}_{\mathrm{m}} := \begin{bmatrix} \hat{C}_1 \\ \hat{C}_2 \\ \vdots \\ \hat{C}_\nu \end{bmatrix}.$$

Define the following information flow structure:

$$\mathbf{K} := \{K \in R^{m\times r} \mid K = \begin{bmatrix} \overline{K}_1 \; K_1 & & & \mathbf{0} \\ & \overline{K}_2 \; K_2 & & \\ & & \ddots & \\ \mathbf{0} & & & \overline{K}_\nu \; K_\nu \end{bmatrix},$$

$$\overline{K}_i \in R^{m_i \times r_i^{\mathrm{m}}},\; K_i \in R^{m_i\times r_i},\; i = 1, \ldots, \nu\}.$$

Lemma 3.2 [1] *A necessary and sufficient condition for the existence of a robust decentralized LTI controller for* (3.1) *such that the error* $e(t)$ *given by* (3.2) *approaches zero as* $t \to \infty$ *for all measurable or unmeasurable disturbances described by* (3.4), *for all specified* $y_{\text{ref}}(t)$ *described by* (3.5), *and such that the closed-loop system is stable is that the following conditions all hold:*

(1) *The triple* (C_{m}, A, B) *has no unstable DFM with respect to* **K**.
(2) *The DFMs with respect to* **K** *of the* q *systems:*

$$\left\{\hat{C}_{\text{m}}, \begin{bmatrix} A & 0 \\ C & \lambda_j I \end{bmatrix}, \begin{bmatrix} B \\ D \end{bmatrix}\right\}, \quad j = 1, \ldots, q$$

do not contain λ_j, $j = 1, \ldots, q$, *respectively.*
(3) $y_i^{\text{m}}(t)$, $i = 1, \ldots, \nu$, *must contain the output* $y_i(t)$, $i = 1, z \ldots, \nu$, *respectively, i.e.* $y_i(t)$, $i = 1, \ldots, \nu$, *can be measured.*

Control structure [1]: Assume that conditions (1)–(3) given in Lemma 3.2 are satisfied; then, any robust decentralized controller which regulates (3.1) must have the following structure:

$$u_i(t) = \hat{K}_i \eta_i(t) + Q_i z_i(t)$$

In the above equation, $z_i(t) \in R^{r_i \cdot q}$ is the output of the decentralized servocompensator given by:

$$\dot{z}_i(t) = S_i^* z_i(t) + R_i^* e_i(t), \quad i = 1, \ldots, \nu$$

where

$$S_i^* = \underbrace{\begin{bmatrix} P & & \mathbf{0} \\ & \ddots & \\ \mathbf{0} & & P \end{bmatrix}}_{r_i \text{ blocks}},$$

and $R_i^* \in R^{r_i \cdot q \times r_i}$ is a real matrix of rank r_i with the property that (S_i^*, R_i^*) is controllable for $i = 1, \ldots, \nu$. $\eta_i(t)$ is the output of a decentralized stabilizing compensator with inputs $\hat{y}_i^{\text{m}}(t) = C_i^{\text{m}} x(t)$ and $z_i(t)$, where Q_i, $\hat{K}_i$ and the parameters of the stabilizing decentralized compensator are found to stabilize the following system:

$$\begin{bmatrix} \dot{x}(t) \\ \dot{z}_1(t) \\ \vdots \\ \dot{z}_\nu(t) \end{bmatrix} = \begin{bmatrix} A & 0 & \cdots & 0 \\ R_1^* C_1 & S_1^* & \cdots & 0 \\ \vdots & \vdots & & \vdots \\ R_\nu^* C_\nu & 0 & \cdots & S_\nu^* \end{bmatrix} \begin{bmatrix} x(t) \\ z_1(t) \\ \vdots \\ z_\nu(t) \end{bmatrix}$$

$$+ \begin{bmatrix} B \\ \text{block diag}\left(\left[R_1^* D_1 \cdots R_\nu^* D_\nu \right] \right) \end{bmatrix} \begin{bmatrix} u_1(t) \\ \vdots \\ u_\nu(t) \end{bmatrix}$$

$$\begin{bmatrix} \hat{y}_i^{\mathrm{m}}(t) \\ z_i(t) \end{bmatrix} = \begin{bmatrix} C_i^{\mathrm{m}} x(t) \\ z_i(t) \end{bmatrix}, \quad i = 1, \ldots, \nu$$

and where the system has no DFMs with respect to **K**.

3.4 Summary

In this chapter, a technique is presented to design a compensator for an LTI system to solve the robust servomechanism problem in both centralized and decentralized forms. Robust servomechanism problem is a fundamental problem in control theory, where the objective is to design a controller so that the output signal tracks a reference input and rejects a disturbance input with no steady-state error when the parameters of the system are subject to change. It is assumed that the reference and disturbance inputs can be expressed as a combination of non-decaying exponentials, sinusoids, ramp signals as well as constant signals (which are commonly used test signals in control). The structure of the controller in both centralized and decentralized forms depends on the characteristics of the reference and disturbance signals.

3.5 Problems

1. Consider the 4-input, 4-output furnace problem (given in furnace.mat). Assume the reference input and disturbance input are all constant signals. Check if there is a solution for:
   (a) The robust centralized servomechanism problem.
   (b) The robust decentralized servomechanism problem.
2. Consider the following system:

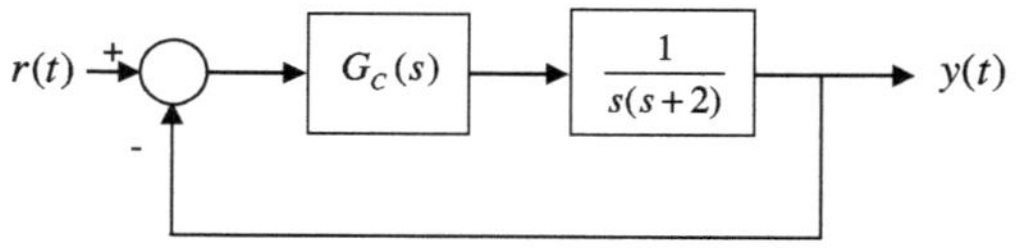

(a) Let the controller $G_C(s)$ be equal to -1. Can the system track constant reference signals $r(t)$ with no steady-state error? Does this controller solve the robust servomechanism problem?

(b) Let the controller $G_C(s)$ be equal to 1. Can the system track constant reference signals $r(t)$ with no steady-state error? Does this controller solve the robust servomechanism problem?

Assume that both plant and controller are controllable and observable.

3. Consider two identical pendulums which are coupled by a spring as follows:

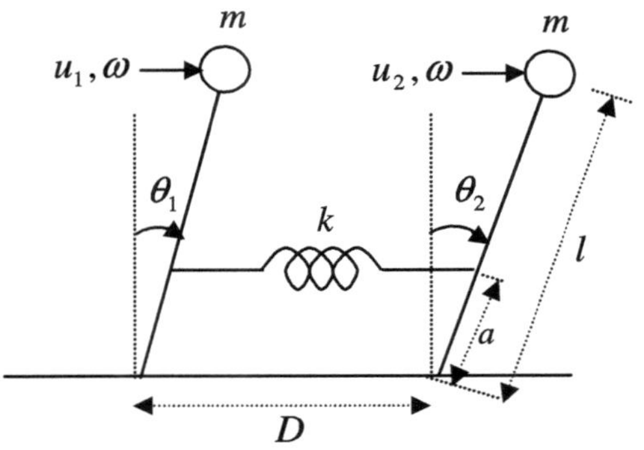

where the spring is linear and its nominal length corresponds to the case $\theta_1 = \theta_2$ (i.e. in this situation no force is applied to the pendulums through the spring). This is, in fact, the model for a robot arm subject to two distinct input torques $u_1(t)$ and $u_2(t)$. Assume that the state vector is $x(t) = \left[\theta_1(t)\ \dot{\theta}_1(t)\ \theta_2(t)\ \dot{\theta}_2(t)\right]'$ and the output signals (which are also the measured signals) are $y_1(t) = \theta_1(t) + 0.1\dot{\theta}_1(t)$ and $y_2(t) = \theta_2(t) + 0.1\dot{\theta}_2(t)$. It is desired to design a decentralized controller of the form:

$$P_i : \begin{cases} \dot{z}_i(t) = S_i z_i(t) + R_i\big(y_i(t) - y_i^{\text{ref}}(t)\big) \\ u_i(t) = Q_i z_i(t) + K_i y_i(t) \end{cases}, i = 1,\ 2$$

to solve the robust decentralized servomechanism problem for this system, where $z_i(t) \in R$ is the state vector of the ith feedback controller, $y_i^{\text{ref}}(t)$ is the constant reference signal of the ith control agent and $\omega(t)$ is the constant disturbance torque. The linearized model for this system is given by:

$$\dot{x}(t) = \begin{bmatrix} 0 & 1 & 0 & 0 \\ \frac{g}{l} - \frac{ka^2}{ml^2} & 0 & \frac{ka^2}{ml^2} & 0 \\ 0 & 0 & 0 & 1 \\ \frac{ka^2}{ml^2} & 0 & \frac{g}{l} - \frac{ka^2}{ml^2} & 0 \end{bmatrix} x(t) + \begin{bmatrix} 0 \\ \frac{1}{ml^2} \\ 0 \\ 0 \end{bmatrix} u_1(t) + \begin{bmatrix} 0 \\ 0 \\ 0 \\ \frac{1}{ml^2} \end{bmatrix} u_2(t) + \begin{bmatrix} 0 \\ \frac{1}{ml^2} \\ 0 \\ \frac{1}{ml^2} \end{bmatrix} \omega(t)$$

$$y_1(t) = \begin{bmatrix} 1 & 0.1 & 0 & 0 \end{bmatrix} x(t)$$

$$y_2(t) = \begin{bmatrix} 0 & 0 & 1 & 0.1 \end{bmatrix} x(t)$$

Verify the existence of a solution for the robust decentralized servomechanism problem for the linear model with $g = 10$, $l = 10$, $m = 0.01$, $k = 0.02$, $a = 5$, and design such a controller if the existence conditions are satisfied.

References

1. E.J. Davison, The robust decentralized control of a general servomechanism problem. IEEE Trans. Autom. Control **AC-21**, 14–24 (1976)
2. W.S. Levine, M. Athans, On the determination of the optimal constant output feedback gains for linear multivariable systems. IEEE Trans. Autom. Control **AC-15**, 44–48 (1970)

Chapter 4
On Tuning Regulators

Up to this point in the book, we have assumed that the plant model has been given, and the objective is to carry out system analysis and controller design. In many situations, however, a good model is not available. The two typical solutions to this problem are

- use physical laws to ascertain a plant model, and then carry out experiments to ascertain parameter values, after which we proceed as before, and
- use adaptive control, wherein one typically carries out recursive estimation and control, learning about the plant and controlling it simultaneously.

In both cases, certain plant structural information is typically needed, such as the plant order, which may be available in simple cases but not readily so in more complicated situations. This brings us to a third, more direct approach, wherein the goal is to design a controller for the plant without the intermediate step of obtaining a detailed model; it is a generalization of the classical approach to tuning a three-term PID controller used for single-input, single-output systems. For this to work, it is typically assumed that

- the plant is (asymptotically) stable;
- it is possible to carry out a finite number of experiments on the plant;
- it is possible to carry out online tuning of controller parameter(s); and
- the goal is to track steps and/or a finite number of sinusoids of known frequency.

The approach has the nice feature that it yields a robust controller: one which ensures the asymptotic tracking feature as long as closed-loop stability is maintained.

For the sake of clarity, in this chapter, we will focus on step tracking. In Sect. 4.1, we provide the mathematical set-up. In Sect. 4.2, we present some necessary conditions for the problem to be solvable. In Sect. 4.3, we consider the centralized version of the problem, while in Sect. 4.4, we consider the decentralized version of the problem. In Sect. 4.5, we present an illustrative example.

E. J. Davison et al., *Decentralized Control of Large-Scale Systems*,
https://doi.org/10.1007/978-1-4419-6014-6_4

4.1 The Mathematical Set-up

In this chapter, we consider the plant model given by

$$\dot{x} = Ax + \sum_{i=1}^{\nu} B_i u_i + Ew \tag{4.1}$$

$$e_i = \underbrace{C_i x + \sum_{i=1}^{\nu} D_{ij} u_j + F_i w}_{=:y_i} - (y_{ref})_i, \quad i = 1, \ldots, \nu, \tag{4.2}$$

with $x(t) \in \mathbf{R}^n$ the plant state, $w(t) \in \mathbf{R}^q$ the (unmeasureable) disturbance input, $u_i(t) \in \mathbf{R}^{m_i}$ the ith control input, $y_i(t) \in \mathbf{R}^{r_i}$ the ith plant output, $(y_{ref})_i \in \mathbf{R}^{r_i}$ the ith reference signal input, and $e_i(t) \in \mathbf{R}^{r_i}$ the ith (tracking) error signal. To form the centralized version of the plant model, first define the matrices

$$B := \begin{bmatrix} B_1 & \cdots & B_\nu \end{bmatrix}, \; C := \begin{bmatrix} C_1 \\ \vdots \\ C_\nu \end{bmatrix}, \; D := \begin{bmatrix} D_{11} & \cdots & D_{1\nu} \\ \vdots & \vdots & \vdots \\ D_{\nu 1} & \cdots & D_{\nu\nu} \end{bmatrix}, \; F := \begin{bmatrix} F_1 \\ \vdots \\ F_\nu \end{bmatrix},$$

then define the vectors

$$e := \begin{bmatrix} e_1 \\ \vdots \\ e_\nu \end{bmatrix}, \; u := \begin{bmatrix} u_1 \\ \vdots \\ u_\nu \end{bmatrix}, \; y_{ref} := \begin{bmatrix} (y_{ref})_1 \\ \vdots \\ (y_{ref})_\nu \end{bmatrix},$$

and finally consider the centralized model

$$\dot{x} = Ax + Bu + Ew \tag{4.3}$$

$$e = \underbrace{Cx + Du + Fw}_{=:y} - y_{ref}; \tag{4.4}$$

we set $m = \sum_{i=1}^{\nu} m_i$ and $r = \sum_{i=1}^{\nu} r_i$. Here, we assume that the plant is stable, i.e. the matrix A is stable, but the values of the matrices A, B, C, D, E and F are unknown; only the plant output y and the reference signal y_{ref} can be measured. The goal is to design a controller which provides closed-loop stability and provides asymptotic tracking of all constant reference signals $y_{ref}(t)$ as well as reject all constant disturbance signals $w(t)$. To achieve this, we are allowed to carry out a finite number of experiments on the plant as well as carry out some online tuning. We first consider this problem in the centralized case, and then consider the more complicated decentralized counterpart. It turns out that the transfer function from u to e, which we label $G(s)$, plays an important role: it is given by

$$G(s) = D + C(sI - A)^{-1}B.$$

4.2 Necessary Conditions and Their Implications

It is well known in classical control that in order to track a step reference signal using an LTI controller which provides closed-loop stability, then it is necessary and sufficient that the plant not have a zero at zero. It turns out that we can prove something much more general, in which we do not impose structure on the form of the controller:

Proposition 4.1 (Lemma A.1 of [11]) *Suppose the stable plant* (4.3)–(4.4) *has the property that for every constant reference signal y_{ref} and every constant disturbance signal w, there exists a bounded control signal u so that x is bounded and the error signal $e(t) \to 0$ as $t \to \infty$. Then the rank of the DC gain matrix $G(0) = D - CA^{-1}B$ equals r.*

Remark 4.1 The condition that the rank of the DC gain matrix $G(0)$ equals r means, in classical terms, that the system has at least as many inputs as outputs and has no transmission zeros at zero. Equivalently, the rank of the matrix

$$\begin{bmatrix} A & B \\ C & D \end{bmatrix}$$

equals $n + r$.

It turns out that one can obtain an estimate of $G(0)$ in the following way. Assuming that the disturbance w and reference input y_{ref} are turned off, suppose that we set

$$u(t) = \bar{u}^1;$$

because A is stable, we see that

$$e(t) \to \; G(0)\bar{u}^1 \; =: \bar{e}^1,$$

regardless of the value of the initial condition $x(0)$. Of course, we cannot wait for an infinite amount of time, so we only obtain an estimate $\hat{\bar{e}}^1$ of $\bar{e}^1$. This procedure can clearly be repeated for m linearly independent $\bar{u}^i$, $i = 1, \ldots, m$, yielding m estimates $\hat{\bar{e}}^i$, $i = 1, \ldots, m$. Hence, we have that

$$\begin{bmatrix} \hat{\bar{e}}^1 & \cdots & \hat{\bar{e}}^m \end{bmatrix} \approx G(0) \begin{bmatrix} \bar{u}^1 & \cdots & \bar{u}^m \end{bmatrix},$$

so an estimate of $G(0)$ is given by

$$\hat{G}(0) := \begin{bmatrix} \hat{\bar{e}}^1 & \cdots & \hat{\bar{e}}^m \end{bmatrix} \begin{bmatrix} \hat{\bar{u}}^1 & \cdots & \hat{\bar{u}}^m \end{bmatrix}^{-1}.$$

It turns out that if this estimate is sufficiently accurate (this will be precisely defined in the next section), then a controller can be designed which will achieve our objectives.

4.3 Controller Design: The Centralized Case

Motivated by Proposition 4.1, at this point we assume that the rank of $G(0)$ equals r. The proposed control law is an integral-based controller of the form

$$\dot{z} = \varepsilon e \tag{4.5}$$

$$u = Kz. \tag{4.6}$$

When this controller is applied to the plant (4.3)–(4.4), we end up with

$$\begin{bmatrix} \dot{x} \\ \dot{z} \end{bmatrix} = \underbrace{\begin{bmatrix} A & BK \\ \varepsilon C & \varepsilon DK \end{bmatrix}}_{=:A_{cl}(\varepsilon,K)} \begin{bmatrix} x \\ z \end{bmatrix} + \begin{bmatrix} E \\ \varepsilon F \end{bmatrix} w + \begin{bmatrix} 0 \\ -\varepsilon I \end{bmatrix} y_{ref}; \tag{4.7}$$

the associated feedback diagram is given in Fig. 4.1. It turns out that if the closed-loop system is stable, then we automatically obtain step tracking and step disturbance rejection:

Theorem 4.1 *If $\varepsilon \neq 0$ and $K \in \mathbf{R}^{m\times r}$ is such that $A_{cl}(\varepsilon, K)$ is stable, then for every $x(0) \in \mathbf{R}^n$, $z(0) \in \mathbf{R}^r$, constant reference signal y_{ref} and constant disturbance signal w, we have that*

$$e(t) \to 0 \text{ as } t \to \infty.$$

Proof Suppose that $A_{cl}(\varepsilon, K)$ is stable, and let $x(0) \in \mathbf{R}^n$, $z(0) \in \mathbf{R}^r$, the constant reference signal y_{ref} and the constant disturbance signal w be arbitrary. Using basic linear systems theory, we know that $\begin{bmatrix} x \\ z \end{bmatrix}(t)$ must be a weighted vector sum

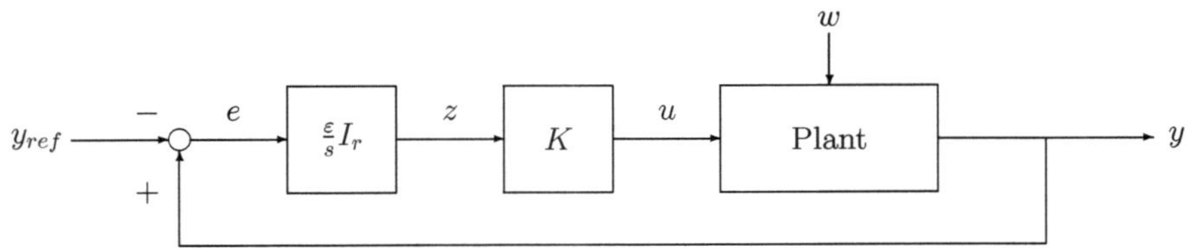

Fig. 4.1 The centralized tuning regulator feedback diagram

of decaying exponentials together with a constant vector; it follows immediately that

$$\lim_{t\to\infty} \begin{bmatrix} \dot{x}(t) \\ \dot{z}(t) \end{bmatrix} = 0.$$

But

$$\dot{z}(t) = \varepsilon e(t),$$

so this means that

$$\lim_{t\to\infty} e(t) = 0$$

as well.

□

So the problem becomes that of choosing ε and K so that $A_{cl}(\varepsilon, K)$ is stable. Now consider carrying out a similarity transformation on this matrix in such a way that $G(0) = D - CA^{-1}B$ appears:

$$\begin{bmatrix} I & A^{-1}BK \\ 0 & I \end{bmatrix} \begin{bmatrix} A & BK \\ \varepsilon C & \varepsilon DK \end{bmatrix} \begin{bmatrix} I & -A^{-1}BK \\ 0 & I \end{bmatrix} = \begin{bmatrix} A + \varepsilon A^{-1}BKC & \varepsilon A^{-1}BKG(0)K \\ \varepsilon C & \varepsilon G(0)K \end{bmatrix}$$
$$=: \bar{A}_{cl}(\varepsilon, K).$$

For more insight, now do a second transformation as follows:

$$\begin{bmatrix} I & 0 \\ 0 & \varepsilon^{1/2}I \end{bmatrix} \bar{A}_{cl}(\varepsilon, K) \begin{bmatrix} I & 0 \\ 0 & \varepsilon^{-1/2}I \end{bmatrix} = \begin{bmatrix} A + \varepsilon A^{-1}BKC & \varepsilon^{1/2}A^{-1}BKG(0)K \\ \varepsilon^{3/2}C & \varepsilon G(0)K \end{bmatrix};$$

if $A + \varepsilon A^{-1}BKC$ and $G(0)K$ happen to be diagonal, we can use Gershgorin circles [9] to conclude that $A_{cl}(\varepsilon, K)$ is stable as long as

- $G(0)K$ is stable, and
- $\varepsilon > 0$ is sufficiently small.

Of course, typically neither are diagonal, so proving this result in the general case requires a bit more work, which we will now proceed to do.

Proposition 4.2 *If $G(0)K$ is stable and $\varepsilon > 0$ is sufficiently small, then $A_{cl}(\varepsilon, K)$ is stable.*

Proof Suppose that $G(0)K$ is stable. Since A is stable as well, there exist square matrices P_1 and P_2 which are symmetric and positive definite so that

$$A^T P_1 + P_1 A = -I$$

and

$$[G(0)K]^T P_2 + P_2[G(0)K] = -I.$$

Now consider the symmetric positive definite matrix

$$P := \begin{bmatrix} P_1 & 0 \\ 0 & P_2 \end{bmatrix};$$

if we can prove that

$$\bar{A}_{cl}(\varepsilon, K)^T P + P\bar{A}_{cl}(\varepsilon, K) \tag{4.8}$$

is negative definite for small $\varepsilon > 0$, then we can conclude from standard Lyapunov theory that $\bar{A}_{cl}(\varepsilon, K)$ is stable (so therefore $A_{cl}(\varepsilon, K)$ is stable as well) for that range of ε. To proceed, first observe that

$$\bar{A}_{cl}(\varepsilon, K)^T P + P\bar{A}_{cl}(\varepsilon, K) = \begin{bmatrix} -I + \varepsilon[(A^{-1}BKC)^T P_1 + P_1(A^{-1}BKC)] & \varepsilon(C^T P_2 + P_1 A^{-1}BKG(0)K) \\ \varepsilon(C^T P_2 + P_1 A^{-1}BKG(0)K)^T & -\varepsilon I \end{bmatrix}.$$

This means that

$$-\frac{1}{\varepsilon}[\bar{A}_{cl}(\varepsilon, K)^T P + P\bar{A}_{cl}(\varepsilon, K)] = \begin{bmatrix} \frac{I}{\varepsilon} - [(A^{-1}BKC)^T P_1 + P_1(A^{-1}BKC)] & -(C^T P_2 + P_1 A^{-1}BKG(0)K) \\ -(C^T P_2 + P_1 A^{-1}BKG(0)K)^T & I \end{bmatrix}.$$

By looking at the principal leading minors, it is easy to see that they are all positive if $\varepsilon > 0$ is small enough, which means that this matrix is positive definite; this, in turn, means that (4.8) is negative definite if $\varepsilon > 0$ is small enough, as desired. □

To apply this result, we proceed as follows. If we know $G(0)$ exactly, then we can set

$$K = -G(0)^T[G(0)G(0)^T]^{-1},$$

in which case $G(0)K = -I$ which is stable; at this point, all we need to do is tune the variable ε. Of course, usually all we have is an estimate of $\hat{G}(0)$ of $G(0)$: assuming that it has full row rank, we can set

$$K = -\hat{G}(0)^T[\hat{G}(0)\hat{G}(0)^T]^{-1};$$

if the estimation error is not too large, then it is clear that $G(0)K$ will be stable so that the same approach works.

Example 4.1 Consider the single-input, single-output system

$$\dot{x} = \begin{bmatrix} 0 & 1 \\ -2 & -3 \end{bmatrix} x + \begin{bmatrix} 0 \\ 1 \end{bmatrix} u$$
$$e = \begin{bmatrix} -1 & 1 \end{bmatrix} x - y_{ref}$$

with a transfer function from u to e of $\frac{s-1}{s^2+3s+2}$. If we set

$$K = -G(0)^{-1} = 2$$

and

$$\dot{z} = \varepsilon e,$$

then we can determine closed-loop stability by examining the transfer function from y_{ref} to e:

$$\frac{-s(s^2+3s+2)}{s(s^2+3s+2) - 2\varepsilon(s-1)} = \frac{-s(s^2+3s+2)}{s^3+3s^2+(2-2\varepsilon)+2\varepsilon}.$$

Applying the Routh-Hurwitz criterion, we conclude that we have closed-loop stability if and only if $\varepsilon \in (0, 0.75)$.

Example 4.2 Consider the multivariable system

$$\dot{x} = \begin{bmatrix} -1 & 0 \\ 0 & -1 \end{bmatrix} x + \begin{bmatrix} 1 & 0 \\ 0 & 1 \end{bmatrix} u$$
$$e = \begin{bmatrix} 0 & 1 \\ 1 & 0 \end{bmatrix} x - y_{ref}$$

If we set

$$K = -G(0)^{-1} = \begin{bmatrix} 0 & -1 \\ -1 & 0 \end{bmatrix}$$

and

$$\dot{z} = \varepsilon[y - y_{ref}],$$

then we can determine closed-loop stability by examining the closed-loop state-space equation

$$\begin{bmatrix} \dot{x} \\ \dot{z} \end{bmatrix} = \begin{bmatrix} A & BK \\ \varepsilon C & \varepsilon DK \end{bmatrix} \begin{bmatrix} x \\ z \end{bmatrix} + \begin{bmatrix} 0 \\ -\varepsilon I \end{bmatrix} y_{ref}$$
$$= \underbrace{\begin{bmatrix} -1 & 0 & 0 & -1 \\ 0 & -1 & -1 & 0 \\ 0 & \varepsilon & 0 & 0 \\ \varepsilon & 0 & 0 & 0 \end{bmatrix}}_{=:A_{cl}(\varepsilon,K)} \begin{bmatrix} x \\ z \end{bmatrix} + \begin{bmatrix} 0 \\ -\varepsilon I \end{bmatrix} y_{ref}.$$

The characteristic polynomial is given by

$$\begin{aligned}\det[sI - A_{cl(\varepsilon,K)}] &= \det\begin{bmatrix} s+1 & 0 & 0 & 1 \\ 0 & s+1 & 1 & 0 \\ 0 & -\varepsilon & s & 0 \\ -\varepsilon & 0 & 0 & s \end{bmatrix} \\ &= (s+1)[s^2(s+1)+\varepsilon s] - [-\varepsilon s(s+1) - \varepsilon^2] \\ &= s^2(s+1)^2 + 2\varepsilon s(s+1) + \varepsilon^2 \\ &= s^4 + 2s^3 + (2\varepsilon+1)s^2 + 2\varepsilon s + \varepsilon^2.\end{aligned}$$

Applying the Routh-Hurwitz criterion, we conclude that we have closed-loop stability if and only if $\varepsilon > 0$.

4.4 Controller Design: The Decentralized Case

In the decentralized case, we have constraints on information flow. Obviously, the necessary condition proven in Proposition 4.1 is still required:

$$\text{rank}[G(0)] = \text{rank}[D - CA^{-1}B] = r.$$

Of course, now this condition may not be sufficient. To proceed, let us apply the natural decentralized counterpart of the integral-based controller:

$$\begin{aligned}\dot{z}_i &= \varepsilon_i e_i \\ u_i &= K_i z_i, \ \ i = 1, \dots, \nu;\end{aligned}$$

the closed-loop feedback configuration is given in Fig. 4.2. To write down this controller in a concise way, we need some notation:

$$z := \begin{bmatrix} z_1 \\ \vdots \\ z_\nu \end{bmatrix}, \ K = \text{diag}\{K_1, \dots, K_\nu\}, \ \mathcal{E} := \text{diag}\{\varepsilon_1 I_{r_1}, \dots, \varepsilon_\nu I_{r_\nu}\},$$

after which we can write

$$\begin{aligned}\dot{z} &= \mathcal{E} e \\ u &= Kz.\end{aligned}$$

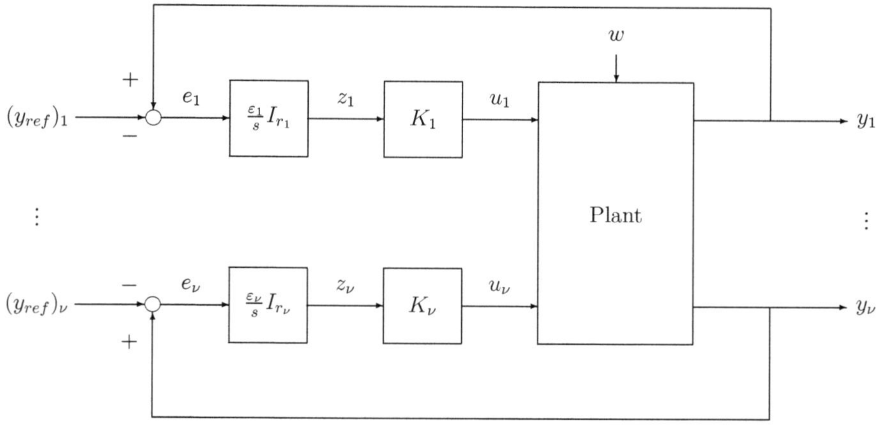

Fig. 4.2 The decentralized tuning regulator feedback diagram

In closed loop, we have

$$\begin{bmatrix} \dot{x} \\ \dot{z} \end{bmatrix} = \underbrace{\begin{bmatrix} A & BK \\ \mathcal{E}C & \mathcal{E}DK \end{bmatrix}}_{=:A_{cl}(\mathcal{E},K)} \begin{bmatrix} x \\ z \end{bmatrix} + \begin{bmatrix} E \\ \mathcal{E}F \end{bmatrix} w + \begin{bmatrix} 0 \\ -\mathcal{E} \end{bmatrix} y_{ref}; \tag{4.9}$$

so for closed-loop stability, we need the following matrix to be stable:

$$A_{cl}(\mathcal{E}, K) = \begin{bmatrix} A & BK \\ \mathcal{E}C & \mathcal{E}DK \end{bmatrix}.$$

So if $\varepsilon_1 = \cdots = \varepsilon_\nu$ are positive and small enough, then by mimicing the centralized argument we see that we achieve the objective as long as

$$G(0)K$$

is stable. Of course, there may be some choices of $G(0)$ of full row rank for which no block diagonal gain K will suffice.

Example 4.3 Consider the case of $p = 2$ in which

$$G(0) = \begin{bmatrix} 0 & g_{12} \\ g_{21} & 0 \end{bmatrix}.$$

Then for a diagonal gain matrix

$$K = \begin{bmatrix} k_1 & 0 \\ 0 & k_2 \end{bmatrix},$$

we have

$$G(0)K = \begin{bmatrix} 0 & g_{12}k_2 \\ g_{21}k_1 & 0 \end{bmatrix},$$

whose characteristic polynomial is

$$s^2 - g_{12}g_{21}k_1k_2;$$

this never has both roots in $\mathbf{C}^-$, i.e. $G(0)K$ is never stable for any choice of diagonal K.

So we conclude that if this form of controller is to be used, then there are some additional constraints needed for this methodology to work. Another issue that we have sidestepped is that of the tuning of the values of ε_i; tuning them together violates the spirit of the decentralized setting. Furthermore, it is natural to wish to tune them in sequence, without retuning ones which have already been tuned. Last of all, it is reasonable to require that we maintain stability at each step.

To formalize this idea, we impose a **decentralized controller tuning synthesis constraint** [3]:

(i) A controller can be tuned on only one control channel at a time.
(i) After a controller has been tuned on a given channel, the controller is fixed and will not be adjusted further.
(iii) After each of the ν tuning procedures, the resulting closed-loop system is stable.

The motivation for (iii) is obvious. The motivation for conditions (i) and (ii) is that these are desirable to have in practise.

Before stating the main result, we consider an illustrative example.

Example 4.4 Consider a system with two scalar channels and no disturbance: in transfer function form, we have

$$E(s) = \begin{bmatrix} E_1(s) \\ E_2(s) \end{bmatrix} = \underbrace{\begin{bmatrix} g_{11}(s) & g_{12}(s) \\ g_{21}(s) & g_{22}(s) \end{bmatrix}}_{=G(s)} \begin{bmatrix} U_1(s) \\ U_2(s) \end{bmatrix} - \begin{bmatrix} Y_{ref,1}(s) \\ Y_{ref,2}(s) \end{bmatrix},$$

and assume that the necessary condition pointed out by Proposition 4.1 holds, i.e. $G(0)$ has rank two. As we see above in Example 4.3, we cannot have both $g_{11}(0) = 0$ and $g_{22}(0) = 0$; for simplicity, assume that $g_{11}(0) \neq 0$. Now choose k_1 so that $g_{11}(0)k_1$ is stable, and then set

$$U_1(s) = \frac{\varepsilon_1}{s} k_1 E_1(s);$$

we know from Proposition 4.2 (applied to the system restricted to channel one) that if $\varepsilon_1 > 0$ is small enough, we have closed-loop stability. Fix such an $\varepsilon_1 > 0$, and now let us analyse the ensuing closed-loop system. With

$$\mathcal{K}_1(s) = \begin{bmatrix} \frac{\varepsilon_1 k_1}{s} & 0 \\ 0 & 0 \end{bmatrix}$$

set

$$U(s) = \mathcal{K}_1(s)E(s) + U^0(s);$$

suppressing the argument in g_{ij} to enhance readability, we end up with

$$\begin{aligned}
E(s) &= [I - G(s)\mathcal{K}_1(s)]^{-1}G(s)U^0(s) - [I - G(s)\mathcal{K}_1(s)]^{-1}Y_{ref}(s) \\
&= \frac{1}{s - \varepsilon_1 k_1 g_{11}} \begin{bmatrix} sg_{11} & sg_{12} \\ sg_{21} & \varepsilon_1 k_1 g_{21} g_{12} + (s - \varepsilon_1 k_1 g_{11}) g_{22} \end{bmatrix} U^0(s) - \\
&\quad \frac{1}{s - \varepsilon_1 k_1 g_{11}} \begin{bmatrix} s & 0 \\ \varepsilon_1 k_1 g_{21} & s - \varepsilon_1 k_1 g_{11} \end{bmatrix} Y_{ref}(s) \\
&=: \begin{bmatrix} h_{11}(s) & h_{12}(s) \\ h_{21}(s) & h_{22}(s) \end{bmatrix} U^0(s) - \frac{1}{s - \varepsilon_1 k_1 g_{11}} \begin{bmatrix} s & 0 \\ \varepsilon_1 k_1 g_{21} & s - \varepsilon_1 k_1 g_{11} \end{bmatrix} Y_{ref}(s);
\end{aligned}$$

closed-loop stability means that the transfer function $\frac{1}{s - \varepsilon_1 k_1 g_{11}}$ is stable. Notice that, if we set $U^0(s) = 0$, then we obtain step tracking in channel one. Of course, we would like to obtain step tracking in channel two as well. To this end, observe that the transfer function from the second element of $U^0(s)$ to the second element of $E(s)$ is

$$h_{22}(s) = \frac{\varepsilon_1 k_1 g_{21}(s) g_{12}(s) + (s - \varepsilon_1 k_1 g_{11}(s)) g_{22}(s)}{s - \varepsilon_1 k_1 g_{11}(s)},$$

which, when evaluated at $s = 0$, yields

$$\frac{\varepsilon_1 k_1 g_{21}(0) g_{12}(0) - \varepsilon_1 k_1 g_{11}(0) g_{22}(0)}{-\varepsilon_1 k_1 g_{11}(0)} = \frac{\det[G(0)]}{g_{11}(0)}.$$

Since we have assumed that $G(0)$ has rank two, this means that $\det[G(0)] \neq 0$. Hence, now choose k_2 so that

$$k_2 h_{22}(0) < 0,$$

or equivalently so that

$$k_2 \frac{\det[G(0)]}{g_{11}(0)} < 0,$$

and set

$$U_2^0(s) = \frac{\varepsilon_2}{s} k_2 E_2(s);$$

we know from the centralized case (applied to the above-compensated system restricted to channel two) that if ε_2 is small enough, then we have closed-loop stability. Fix such an ε_2, and now analyse the ensuing closed-loop system. With

$$\mathcal{K}_2(s) = \begin{bmatrix} 0 & 0 \\ 0 & \frac{\varepsilon_2 k_2}{s} \end{bmatrix}$$

set

$$U^0(s) = \mathcal{K}_2(s)E(s);$$

suppressing the argument in g_{ij} and h_{ij} to enhance readability, we end up with

$$\begin{aligned} E(s) &= -\frac{1}{[s-\varepsilon_2 k_2 h_{22}][s-\varepsilon_1 k_1 g_{11}]} \times \\ &\quad \begin{bmatrix} s-\varepsilon_2 k_2 h_{22} & \varepsilon_2 k_2 h_{12} \\ 0 & s \end{bmatrix} \times \begin{bmatrix} s & 0 \\ \varepsilon_1 k_1 g_{21} & s-\varepsilon_1 k_1 g_{11} \end{bmatrix} Y_{ref}(s) \\ &= -\frac{1}{[s-\varepsilon_2 k_2 h_{22}][s-\varepsilon_1 k_1 g_{11}]} \times \\ &\quad \begin{bmatrix} s(s-\varepsilon_2 k_2 h_{22}) + \varepsilon_1 \varepsilon_2 k_1 k_2 h_{12} g_{21} & \varepsilon_2 k_2 h_{12}(s-\varepsilon_1 k_1 g_{11}) \\ \varepsilon_1 k_1 g_{21} s & s(s-\varepsilon_1 k_1 g_{11}) \end{bmatrix} Y_{ref}(s); \end{aligned}$$

closed-loop stability means that the transfer function $\frac{1}{s-\varepsilon_2 k_2 h_{22}}$ is stable. To achieve zero steady-state tracking error for step reference signals, we need the transfer function from y_{ref} to e to be zero at $s = 0$: using the fact that $h_{12}(0) = 0$, we conclude that this is clearly the case. This example motivates the forthcoming design procedure.

Before presenting the main result, we first impose a regularity condition on D:

Assumption 1: $D_{ij} = 0$ if $i \neq j$.

Just as the DC gain $G(0) = D - CA^{-1}B$ plays a critical role in the centralized case, the quantity

$$T_{ij} := D_{ij} - C_i A^{-1} B_j \in \mathbf{R}^{r_i \times m_i}$$

plays an important role in the decentralized case. The system DC gain $G(0)$ can be found by offline experiments as in the centralized case, and the quantities T_{ij} determined accordingly.

Now, we present the main result; for clarity, we first present the square case in which $m_i = r_i$ for all $i = 1, \ldots, \nu$.

Theorem 4.2 [3] *Suppose that $m_i = r_i$ for $i = 1, \ldots, \nu$ and that Assumption*

1 holds. If there exists an ordering of the set $\{1, 2, \ldots, \nu\}$*, namely* $i_1, \ldots, i_\nu$*, which satisfy*

(i) $det\,[T_{i_1,i_1}] \neq 0$,

(ii) $det \begin{bmatrix} T_{i_1,i_1} & T_{i_1,i_2} \\ T_{i_2,i_1} & T_{i_2,i_2} \end{bmatrix} \neq 0$,

$\vdots$

(ν) $det \begin{bmatrix} T_{i_1,i_1} & \cdots & T_{i_1,i_\nu} \\ \vdots & \vdots & \vdots \\ T_{i_\nu,i_1} & \cdots & T_{i_\nu,i_\nu} \end{bmatrix} \neq 0$,

then there exists a decentralized controller design procedure (which satisfies the decentralized controller tuning synthesis constraint) that yields a controller which provides closed-loop stability as well as asymptotic error regulation for all constant reference and disturbance signals.

Remark 4.2 The design procedure is a natural extension of that observed in the two scalar channel set-up of Example 4.4. More precisely, we follow the following steps:

Step 1: Since $\det[T_{i_1,i_1}] \neq 0$, we can use the centralized theory of Sect. 4.3 to design an integral-based controller in channel i_1 which ensures closed-loop stability and tracking and disturbance rejection of steps in that channel.

Step 2: By carrying out an experiment on the partially controlled plant arising from Step 1, we obtain the new DC gain from input u_{i_2} to output y_{i_2}; this quantity will, in general, be different from that of T_{i_2,i_2}, but due to requirements (i) and (ii), it will always be invertible, even if T_{i_2,i_2} is not. We can now use the centralized theory of Sect. 4.3 to design an integral-based controller in channel i_2 which ensures closed-loop stability and tracking and disturbance rejection of steps in that channel; this does not affect the controller designed in Step 1 for channel i_1.

Step 3: By carrying out an experiment on the partially controlled plant arising from Steps 1 and 2, we obtain the new DC gain from input u_{i_3} to output y_{i_3}; this quantity will, in general, be different from that of T_{i_3,i_3}, but due to requirements (i), (ii) and (iii), it will always be invertible, even if T_{i_3,i_3} is not. We can now use the centralized theory of Sect. 4.3 to design an integral-based controller in channel i_3 which ensures closed-loop stability and tracking and disturbance rejection of steps in that channel; this does not affect the controller designed in Step 1 for channel i_1 nor the controller designed in Step 2 for channel i_2.

Steps 4-ν: We simply repeat the above argument, at each point leaving the controllers designed in the previous steps fixed.

Remark 4.3 If ν is small, we can check the condition of Theorem 4.2 by looking at all of the $\nu!$ permutations of the set $\{1, \ldots, \nu\}$. If ν is too large, this may not be feasible.

Now we turn to the case in which some of the channels are not square. First of all, given that knowledge of the reference signal $(y_{ref})_i$ is local to channel i, it is reasonable at this point to impose

Assumption 2: $m_i \geq r_i$ for every $i = 1, \ldots, \nu$.

The key idea in the non-square case is that for non-square channels we regularize the input by multiplying by a constant matrix of appropriate dimensions which converts it effectively into the square case. To this end, with $K_j \in \mathbf{R}^{m_j \times r_j}$, we define the related quantity

$$T_{ij}(K_j) := T_{ij}K_j = [D_{ij} - C_i A^{-1} B_j]K_j \in \mathbf{R}^{r_i \times r_i}.$$

This brings us to the second main result of this section:

Theorem 4.3 [3] *Suppose that Assumptions 1 and 2 hold. If there exist matrices $K_i \in \mathbf{R}^{m_i \times r_i}$, $i = 1, \ldots, \nu$, as well as an ordering of the set $\{1, 2, \ldots, \nu\}$, namely $i_1, \ldots, i_\nu$, which satisfy*

(i) $det\ [T_{i_1,i_1}(K_{i_1})] \neq 0,$

(ii) $det \begin{bmatrix} T_{i_1,i_1}(K_{i_1}) & T_{i_1,i_2}(K_{i_2}) \\ T_{i_2,i_1}(K_{i_1}) & T_{i_2,i_2}(K_{i_2}) \end{bmatrix} \neq 0,$

$\vdots$

(ν) $det \begin{bmatrix} T_{i_1,i_1}(K_{i_1}) & \cdots & T_{i_1,i_\nu}(K_{i_\nu}) \\ \vdots & \vdots & \vdots \\ T_{i_\nu,i_1}(K_{i_1}) & \cdots & T_{i_\nu,i_\nu}(K_{i_\nu}) \end{bmatrix} \neq 0,$

then there exists a decentralized controller design procedure (which satisfies the decentralized controller tuning synthesis constraint) that yields a controller which provides closed-loop stability as well as asymptotic error regulation for all constant reference and disturbance signals.

Remark 4.4 The procedure in this case is simply an extension of that for the square case considered in Theorem 4.2. With the gains K_i and the ordering of the channels chosen as in Theorem 4.3, simply define

$$u_i = K_i u_i^{new};$$

if we now define $B_i^{new} := B_i K_i$ and regard u_i^{new} as the new ith plant control input, then we are back to the square case, and we can therefore use the algorithm of Remark 4.2.

4.5 A Design Example

Here, we consider a design example to illustrate the ideas presented in this chapter. Consider the plant model

$$\dot{x} = \begin{bmatrix} -2 & 1 & 0 \\ 1 & -3 & 1 \\ 0 & 1 & -4 \end{bmatrix} x + \begin{bmatrix} 1 \\ 1 \\ 0 \end{bmatrix} u_1 + \begin{bmatrix} 0 \\ 1 \\ 1 \end{bmatrix} u_2 + \begin{bmatrix} 0 \\ 0 \\ 1 \end{bmatrix} u_3 + \begin{bmatrix} 1 \\ 0 \\ 1 \end{bmatrix} w$$

$$e_1 = \begin{bmatrix} 1 & 1 & 1 \end{bmatrix} x - \frac{2}{3} u_1 + w - (y_{ref})_1$$

$$e_2 = \begin{bmatrix} 0 & 1 & 0 \end{bmatrix} x - \frac{5}{9} u_2 + 2w - (y_{ref})_2$$

$$e_3 = \begin{bmatrix} 0 & 1 & 1 \end{bmatrix} x - \frac{2}{3} u_3 + 3w - (y_{ref})_3.$$

The underlying assumption in the design procedure is that the model is unknown, but it is allowed to carry out experiments on the plant (in the absence of the disturbance). To this end, suppose that we carry out a sequence of experiments:

1. First of all, set $u_1(t) = 1$ and $u_2(t) = u_3(t) = 0$, and find the "steady-state value" of $e(t)$; of course, we cannot wait an infinite amount of time, so we end up with an estimate:
$$\lim_{t \to \infty} e(t) \approx \begin{bmatrix} 1 \\ 0.7 \\ .8 \end{bmatrix}.$$
2. Next, we set $u_2(t) = 1$ and $u_1(t) = u_3(t) = 0$, and find the "steady-state value" of $e(t)$; we end up with
$$\lim_{t \to \infty} e(t) \approx \begin{bmatrix} 1.2 \\ 0 \\ .9 \end{bmatrix}.$$
3. Next, we set $u_3(t) = 1$ and $u_1(t) = u_2(t) = 0$, and find the "steady-state value" of $e(t)$; we end up with
$$\lim_{t \to \infty} e(t) \approx \begin{bmatrix} 1.3 \\ 0.3 \\ 0 \end{bmatrix}.$$

Hence, our estimate of the DC gain is

$$\hat{G}(0) = \begin{bmatrix} 1 & 1.2 & 1.3 \\ 0.7 & 0 & 0.3 \\ 0.8 & 0.9 & 0 \end{bmatrix}.$$

The control objective is to design a controller which provides closed-loop stability, tracking of constant reference signals, and rejection of constant disturbance signals. To fulfil this objective, we first use the tuning regulator paradigm to first carry out a centralized controller design, and we then carry out a centralized controller design.

4.5.1 A Centralized Controller Design

Following Sect. 4.3, we first confirm that the rank of $\hat{G}(0)$ equals the number of outputs (three). Next, we define

$$K = -\hat{G}(0)^{-1},$$

apply the controller

$$\begin{aligned}\dot{z} &= \varepsilon e\\ u &= Kz,\end{aligned}$$

and then tune the value of ε to provide closed-loop stability and acceptable closed-loop performance. First of all, it is easy to verify that the closed-loop system is stable for $\varepsilon \in (0, 4.77)$. Second of all, we carry out a number of closed-loop experiments on the plant for various values of ε, starting small, and slowly increasing it (since the closed-loop system may go unstable). While there are many ways to tune a single parameter in a practical setting,[1] for the sake of our simulation, we will choose a systematic method so that it can be emulated: we find the value of ε which minimizes a particular performance index. For the case of $w = 0$ and $y_{ref} = \begin{bmatrix} 1 \\ 1 \\ 1 \end{bmatrix}$, in Fig. 4.3, we plot the 2−norm of the error (of the first ten seconds of the response)[2] as a function of ε^3. From this, we see that the optimal ε is approximately $\varepsilon = 1.5$. So for this choice of ε, we plot the step response of the error in Fig. 4.4.

[1]The classical way to tune a variable is to examine the step response and find the one which has the "nicest" step response; the problem is that everyones definition of "nicest" is different.

[2]We represent this by $\|\hat{e}\|_2 := [\int_0^{10} \|e(t)\|_2^2\, dt]^{1/2}$.

[3]In a practical situation, we would only be able to try a small number of choices of ε, many fewer than used here to create Fig. 4.3.

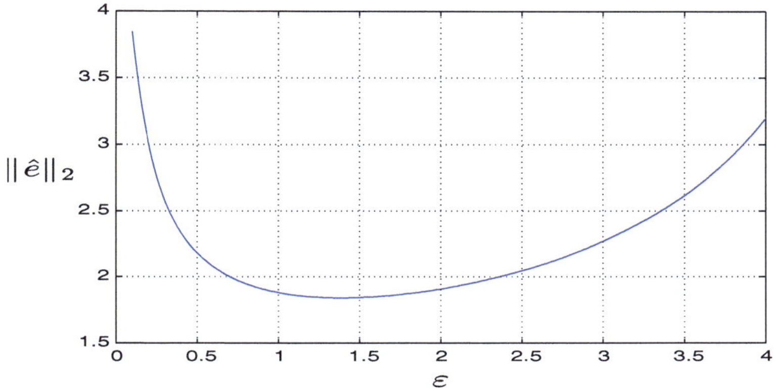

Fig. 4.3 A plot of the 2−norm of the error as a function of ε

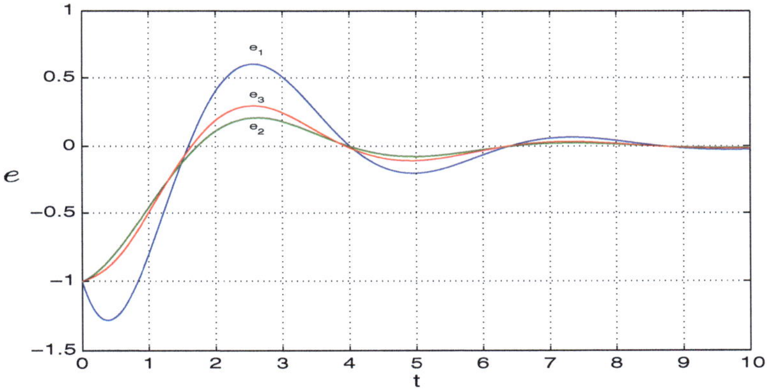

Fig. 4.4 A plot of the error for $\varepsilon = 1.5$

4.5.2 *A Decentralized Controller Design*

Now we would like to apply Theorem 4.2. Of course, we have to work with an estimate $\hat{G}(0)$ rather than $G(0)$:

$$\begin{bmatrix} \hat{T}_{1,1} & \hat{T}_{1,2} & \hat{T}_{1,3} \\ \hat{T}_{2,1} & \hat{T}_{2,2} & \hat{T}_{2,3} \\ \hat{T}_{3,1} & \hat{T}_{3,2} & \hat{T}_{3,3} \end{bmatrix} = \hat{G}(0) = \begin{bmatrix} 1 & 1.2 & 1.3 \\ 0.7 & 0 & 0.3 \\ 0.8 & 0.9 & 0 \end{bmatrix}.$$

We see that we cannot set i_1 equal to 2 or 3 because of the zeros in $\hat{T}_{2,2}$ and $\hat{T}_{3,3}$; however, $i_1 = 1$ will work. If we now set $i_2 = 2$ and $i_3 = 3$, we meet the requirements of Theorem 4.2, at least when using the estimates of the DC gain:

$$\hat{T}_{1,1} = 1 \neq 0,\ \det \begin{bmatrix} \hat{T}_{1,1} & \hat{T}_{1,2} \\ \hat{T}_{2,1} & \hat{T}_{2,2} \end{bmatrix} = \det \begin{bmatrix} 1 & 1.2 \\ 0.7 & 0 \end{bmatrix} = -0.84 \neq 0,$$

$$\det \begin{bmatrix} \hat{T}_{1,1} & \hat{T}_{1,2} & \hat{T}_{1,3} \\ \hat{T}_{2,1} & \hat{T}_{2,2} & \hat{T}_{2,3} \\ \hat{T}_{3,1} & \hat{T}_{3,2} & \hat{T}_{3,3} \end{bmatrix} = \det \begin{bmatrix} 1 & 1.2 & 1.3 \\ 0.7 & 0 & 0.3 \\ 0.8 & 0.9 & 0 \end{bmatrix} = 0.837 \neq 0.$$

To proceed, we follow the algorithm in Remark 4.2.

Step 1: Choose k_1 so that $k_1 T_{1,1}$ is stable[4]; since $\hat{T}_{1,1} = 1$, we will set $k_1 = -1$. Then we apply the controller

$$\begin{aligned} \dot{z}_1 &= \varepsilon_1 e_1 \\ u_1 &= k_1 z_1 = -z_1 \end{aligned}$$

and tune the value of ε_1 to provide closed-loop stability and acceptable closed-loop performance in channel 1. If we focus on the first channel only and set $w = 0$, then the closed-loop system is described by

$$\begin{aligned} \begin{bmatrix} \dot{x} \\ \dot{z}_1 \end{bmatrix} &= \begin{bmatrix} A & B_1 k_1 \\ \varepsilon_1 C_1 & \varepsilon_1 D_{11} k_1 \end{bmatrix} \begin{bmatrix} x \\ z_1 \end{bmatrix} + \begin{bmatrix} 0 \\ -\varepsilon_1 \end{bmatrix} (y_{ref})_1 \\ e_1 &= \begin{bmatrix} C_1 & D_{11} k_1 \end{bmatrix} \begin{bmatrix} x \\ z_1 \end{bmatrix} - (y_{ref})_1 . \end{aligned}$$

First of all, it is easy to verify that the closed-loop system is stable for $\varepsilon \in (0, 1.85)$. Second of all, we carry out a number of closed-loop experiments on the plant for various values of ε_1, starting small, and slowly increasing it (since the closed-loop system may go unstable). For the case of $w = 0$ and $(y_{ref})_1 = 1$, in Fig. 4.5, we plot the 2−norm of the error e_1 (of the first 10 s of the response)[5] as a function of ε_1. From this, we see that the optimal ε_1 is approximately 0.75. So, we choose this value for ε_1, and we plot the step response of the error in channel one in Fig. 4.6. So at this point, we have tuned the controller in channel one.

Step 2: Now we need to tune the controller in channel $i_2 = 2$. Before proceeding, we need to obtain the new DC gain in that channel which arises when the controller from channel 1 is applied: our partially controlled system (ignoring channel 3 and setting the disturbance to zero) is

[4] We use k_1 rather than K_1 to emphasize the fact that it is a scalar.

[5] We represent this by $\|\hat{e}_1\|_2 := [\int_0^{10} e_1(t)^2\, dt]^{1/2}$.

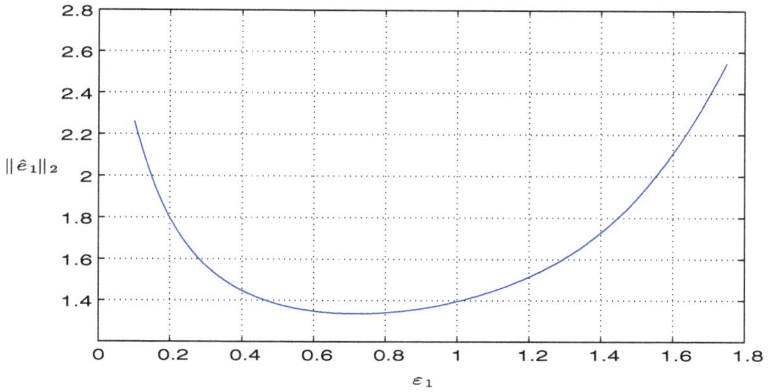

Fig. 4.5 A plot of the 2−norm of the error in channel 1 as a function of ε_1

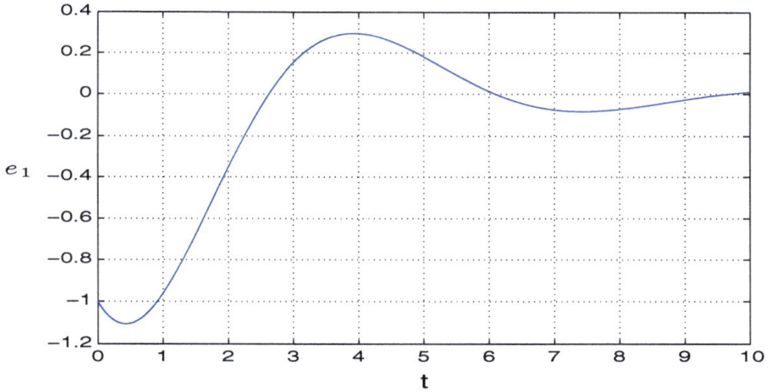

Fig. 4.6 A plot of the error e_1 for $\varepsilon_1 = 0.75$

$$\begin{bmatrix} \dot{x} \\ \dot{z}_1 \end{bmatrix} = \begin{bmatrix} A & B_1 k_1 \\ \varepsilon_1 C_1 & \varepsilon_1 D_{11} k_1 \end{bmatrix} \begin{bmatrix} x \\ z_1 \end{bmatrix} + \begin{bmatrix} 0 \\ -\varepsilon_1 \end{bmatrix} (y_{ref})_1 + \begin{bmatrix} B_2 \\ 0 \end{bmatrix} u_2$$

$$e_1 = \begin{bmatrix} C_1 & D_{11} k_1 \end{bmatrix} \begin{bmatrix} x \\ z_1 \end{bmatrix} - (y_{ref})_1$$

$$e_2 = \begin{bmatrix} C_2 & 0 \end{bmatrix} \begin{bmatrix} x \\ z_1 \end{bmatrix} + D_{22} u_2 - (y_{ref})_2;$$

if we turn off y_{ref} but set $u_2(t) = 1$, then we can obtain an estimate of the "steady-state value" of $e_2(t)$; we end up with

$$\lim_{t \to \infty} e_2(t) \approx -0.8 =: T^*_{2,2};$$

it is not a coincidence that it is approximately equal to

$$\frac{1}{\hat{T}_{1,1}} \det \begin{bmatrix} \hat{T}_{1,1} & \hat{T}_{1,2} \\ \hat{T}_{2,1} & \hat{T}_{2,2} \end{bmatrix}.$$

(They are exactly the same if there is no approximation - see Example 4.4.) So to proceed, we choose k_2 so that $k_2 T^*_{2,2}$ is stable; we will choose $k_2 = 1$. Now we apply the control law

$$\dot{z}_2 = \varepsilon_2 e_2$$
$$u_2 = k_2 z_2$$

and tune the value of ε_2 to provide closed-loop stability and acceptable closed-loop performance. If we focus on channel 2 and set $w = 0$, then the closed-loop system is now described by

$$\begin{bmatrix} \dot{x} \\ \dot{z}_1 \\ \dot{z}_2 \end{bmatrix} = \begin{bmatrix} A & B_1 k_1 & B_2 k_2 \\ \varepsilon_1 C_1 & \varepsilon_1 D_{11} k_1 & 0 \\ \varepsilon_2 C_2 & 0 & \varepsilon_2 D_{22} k_2 \end{bmatrix} \begin{bmatrix} x \\ z_1 \\ z_2 \end{bmatrix} + \begin{bmatrix} 0 \\ 0 \\ -\varepsilon_2 \end{bmatrix} (y_{ref})_2$$

$$e_2 = \begin{bmatrix} C_2 & 0 & D_{22} k_2 \end{bmatrix} \begin{bmatrix} x \\ z_1 \\ z_2 \end{bmatrix} - (y_{ref})_2.$$

First of all, it is easy to verify that the closed-loop system is stable for $\varepsilon_2 \in (0, 1.12)$. Second of all, we carry out a number of closed-loop experiments on the plant for various values of ε_2 starting small, and slowly increasing it (since the closed-loop system may go unstable). For the case of $w = 0$ and $(y_{ref})_2 = 1$, in Fig. 4.7, we plot

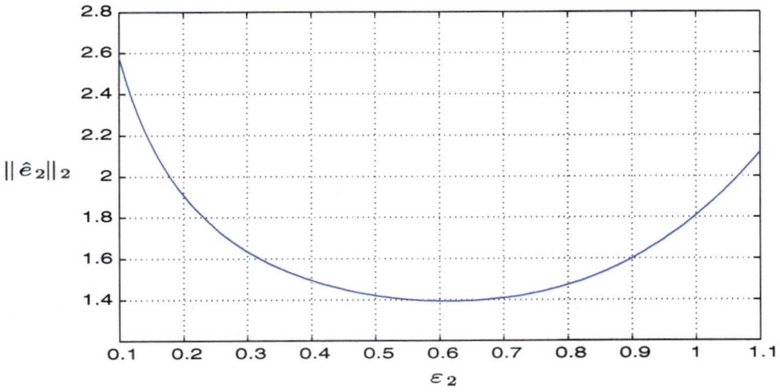

Fig. 4.7 A plot of the 2−norm of the error in channel 2 as a function of ε_2

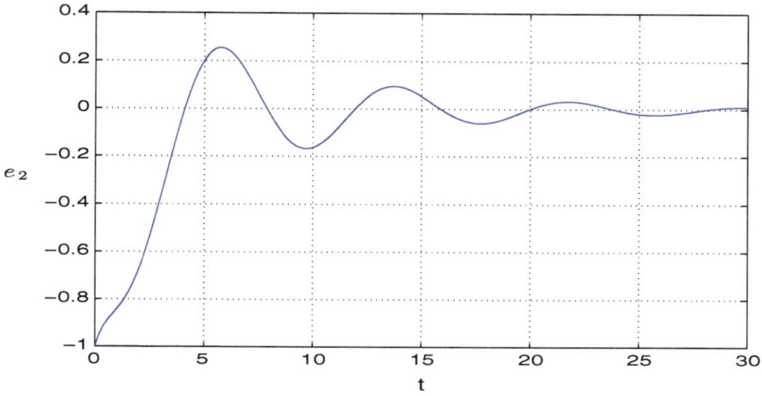

Fig. 4.8 A plot of the error e_2 for $\varepsilon_2 = 0.6$

the 2−norm of the error e_2 (of the first 30 s of the response, since it seems more sluggish than in the previous case)[6] as a function of ε_2. From this, we see that the optimal ε_2 is approximately 0.60. So, we choose this value for ε_2 and we plot the step response of the error in channel two in Fig. 4.8.

Step 3: Now we need to tune the controller in channel $i_3 = 3$. Before proceeding, we need to obtain the new DC gain in that channel which arises when the controller from channels 1 and 2 are applied and $w = 0$: our partially controlled system is

$$\begin{bmatrix} \dot{x} \\ \dot{z}_1 \\ \dot{z}_2 \end{bmatrix} = \begin{bmatrix} A & B_1k_1 & B_2k_2 \\ \varepsilon_1 C_1 & \varepsilon_1 D_{11}k_1 & 0 \\ \varepsilon_2 C_2 & 0 & \varepsilon_2 D_{22}k_2 \end{bmatrix} \begin{bmatrix} x \\ z_1 \\ z_2 \end{bmatrix} + \begin{bmatrix} 0 \\ -\varepsilon_1 \\ 0 \end{bmatrix} (y_{ref})_1 + \begin{bmatrix} 0 \\ 0 \\ -\varepsilon_2 \end{bmatrix} (y_{ref})_2 + \begin{bmatrix} B_3 \\ 0 \\ 0 \end{bmatrix} u_3$$

$$e_1 = \begin{bmatrix} C_1 & D_{11}k_1 & 0 \end{bmatrix} \begin{bmatrix} x \\ z_1 \\ z_2 \end{bmatrix} - (y_{ref})_1$$

$$e_2 = \begin{bmatrix} C_2 & 0 & D_{22}k_2 \end{bmatrix} \begin{bmatrix} x \\ z_1 \\ z_2 \end{bmatrix} - (y_{ref})_2$$

$$e_3 = \begin{bmatrix} C_3 & 0 & 0 \end{bmatrix} \begin{bmatrix} x \\ z_1 \\ z_2 \end{bmatrix} + D_{33}u_3 - (y_{ref})_3;$$

[6] We represent this by $\|\hat{e}_2\|_2 := [\int_0^{30} e_2(t)^2\, dt]^{1/2}$.

if we turn off y_{ref} but set $u_3(t) = 1$, then we can obtain an estimate of the "steady-state value" of $e_3(t)$; we end up with

$$\lim_{t\to\infty} e_3(t) \approx -1.1 =: T^*_{3,3} \neq 0.$$

So to proceed, we choose k_3 so that $k_3 T^*_{3,3}$ is stable; we will choose $k_3 = 1$. Now, we apply the control law

$$\begin{aligned} \dot{z}_3 &= \varepsilon_3 e_3 \\ u_3 &= k_3 z_3 \end{aligned}$$

and tune the value of ε_3 to provide closed-loop stability and acceptable closed-loop performance. If we focus on channel 3 and set $w = 0$, then the closed-loop system is now described by

$$\begin{bmatrix} \dot{x} \\ z_1 \\ z_2 \\ z_3 \end{bmatrix} = \begin{bmatrix} A & B_1 k_1 & B_2 k_2 & B_3 k_3 \\ \varepsilon_1 C_1 & \varepsilon_1 D_{11} K_1 & 0 & 0 \\ \varepsilon_2 C_2 & 0 & \varepsilon_2 D_{22} k_2 & 0 \\ \varepsilon_3 C_3 & 0 & 0 & \varepsilon_3 D_{33} k_3 \end{bmatrix} \begin{bmatrix} x \\ z_1 \\ z_2 \\ z_3 \end{bmatrix} + \begin{bmatrix} 0 \\ 0 \\ 0 \\ -\varepsilon_3 \end{bmatrix} (y_{ref})_3$$

$$e_3 = \begin{bmatrix} C_3 & 0 & 0 & D_{33} k_3 \end{bmatrix} \begin{bmatrix} x \\ z_1 \\ z_2 \\ z_3 \end{bmatrix} - (y_{ref})_3.$$

First of all, it is easy to verify that the closed-loop system is stable for $\varepsilon_3 \in (0, 0.262)$. Second of all, we carry out a number of closed-loop experiments on the plant for various values of ε_3 starting small, and slowly increasing it (since the closed-loop system may go unstable). For the case of $w = 0$ and $(y_{ref})_3 = 1$, in Fig. 4.9, we plot the 2−norm of the error e_3 (of the first 100 s of the response, since it seems more sluggish than in the previous case)[7] as a function of ε_3. From this, we see that the optimal ε_3 is approximately 0.19. So, we choose this value for ε_3 and we plot the step response of the error in channel three in Fig. 4.10. As a last step, we compare our final design to that of the centralized design. We set $y_{ref} = \begin{bmatrix} 1 \\ 1 \\ 1 \end{bmatrix}$ and plot the resulting error signal in Fig. 4.11. Not surprisingly, the response is inferior to the centralized design (see Fig. 4.4). Furthermore, it could very well be that a better overall design can be obtained by designing the three controllers to reduce the overshoot; indeed, if we simply divide each ε_i by two, we get a much better response (Fig. 4.12). This

[7] We represent this by $\|\hat{e}_3\|_2 := [\int_0^{100} e_3(t)^2\, dt]^{1/2}$.

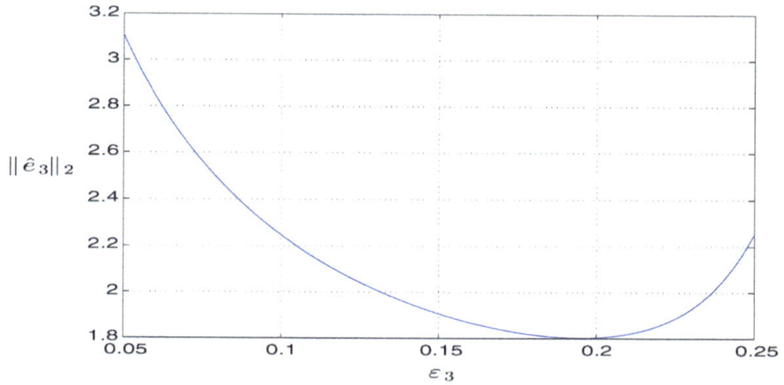

Fig. 4.9 A plot of the 2−norm of the error in channel 3 as a function of ε_3

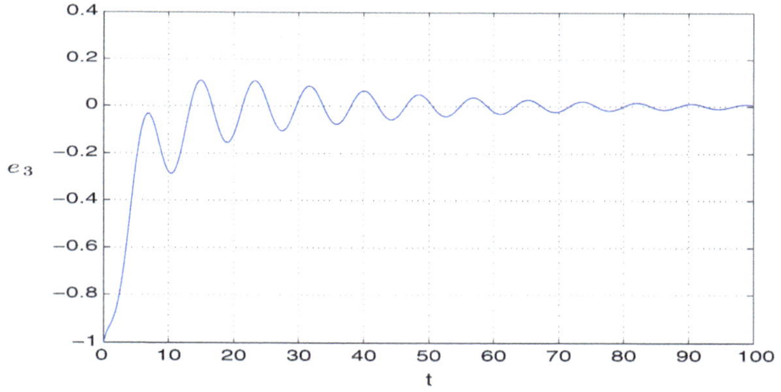

Fig. 4.10 A plot of the error e_3 for $\varepsilon_3 = 0.19$

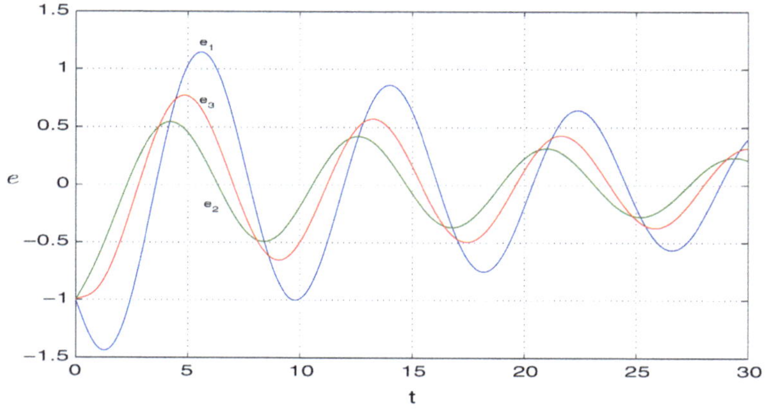

Fig. 4.11 A plot of the error response for the decentralized controller design

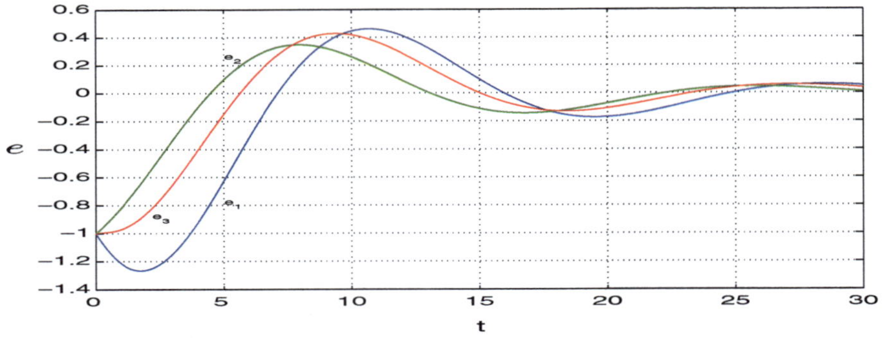

Fig. 4.12 A plot of the error response for the decentralized controller design with each ε_i divided by two

simply emphasizes a feature well known to practitioners: online tuning of controllers is very much an art. That being said, this example demonstrates the details of the proposed approach and shows that it can be used effectively to design step-tracking controllers for plants for which we have no accurate model.

4.6 Summary

The motivation for tuning regulators is that of the classical techniques used to tune PID controllers. The idea was formalized in [2] in the centralized setting, and then extended to the decentralized setting in [3]; in both cases, the goal is to track reference signals (and reject disturbance signals) which are described by a finite sum of sinusoids of known frequency. This idea turned out to be quite useful for a variety of applications, including in the control of distillation columns [1], temperature control of buildings [5], heat exchangers [4], load and power frequency control [6], and microgrids [7, 8], to list but a few. The centralized approach has been extended in a variety of ways, including

(i) the design of PI controllers [13], where it is used to control a concentration-flow process,
(ii) the adaptive setting, in which the controller parameters are tuned automatically [11, 12], and
(iii) the infinite-dimensional setting [10].

4.7 Problems

1. Suppose that $G(0) = \begin{bmatrix} 1 & 0 \\ 0 & 2 \end{bmatrix}$. Characterize all matrices of the form $K = \begin{bmatrix} k_1 & k_2 \\ k_3 & k_4 \end{bmatrix}$ for which $G(0)K$ is stable.

2. For each of the following DC gain matrices, determine if they satisfy the determinant conditions given in Theorem 4.2:

   (a) $G(0) = \begin{bmatrix} 1 & 1 \\ 1 & 1 \end{bmatrix}$.

   (b) $G(0) = \begin{bmatrix} 1 & 1 \\ 1 & 0 \end{bmatrix}$.

   (c) $G(0) = \begin{bmatrix} 0 & 1 \\ 1 & 1 \end{bmatrix}$.

   (d) $G(0) = \begin{bmatrix} 1 & 1 & 1 \\ 1 & 0 & 1 \\ 1 & 1 & 0 \end{bmatrix}$

   (e) $G(0) = \begin{bmatrix} 0 & 1 & 1 \\ 1 & 0 & 1 \\ 1 & 1 & 1 \end{bmatrix}$

   (f) $G(0) = \begin{bmatrix} 0 & 1 & 1 \\ 1 & 0 & 1 \\ 1 & 1 & 2 \end{bmatrix}$

   (g) $G(0) = \begin{bmatrix} \begin{bmatrix} 1 & 1 \\ 1 & 1 \end{bmatrix} & \begin{bmatrix} 1 \\ 0 \end{bmatrix} \\ \begin{bmatrix} 1 & 0 \end{bmatrix} & 0 \end{bmatrix}$

   (h) $G(0) = \begin{bmatrix} \begin{bmatrix} 1 & 1 \\ 1 & 1 \end{bmatrix} & \begin{bmatrix} 1 & 2 \\ 2 & 1 \end{bmatrix} \\ \begin{bmatrix} 1 & 2 \\ 2 & 1 \end{bmatrix} & \begin{bmatrix} 2 & 2 \\ 2 & 2 \end{bmatrix} \end{bmatrix}$

   (i) $G(0) = \begin{bmatrix} 1 & 1 & 0 \\ 1 & 1 & 1 \\ 0 & 1 & 1 \end{bmatrix}$

   (j) $G(0) = \begin{bmatrix} 1 & \begin{bmatrix} 1 & 1 \end{bmatrix} \\ \begin{bmatrix} 1 \\ 0 \end{bmatrix} & \begin{bmatrix} 0 & 1 \\ 0 & 1 \end{bmatrix} \end{bmatrix}$

3. For each of the following DC gain matrices, determine if they satisfy the determinant conditions given in Theorem 4.3:

   (a) $G(0) = \begin{bmatrix} 0 & \begin{bmatrix} 1 & 1 \end{bmatrix} \\ 1 & \begin{bmatrix} 1 & -1 \end{bmatrix} \end{bmatrix}$

   (b) $G(0) = \begin{bmatrix} 0 & \begin{bmatrix} 1 & 1 \end{bmatrix} \\ 0 & \begin{bmatrix} 1 & -1 \end{bmatrix} \end{bmatrix}$

4. Consider the plant

$$\dot{x} = \begin{bmatrix} -1 & 0 & 0 \\ 0 & -2 & 0 \\ 0 & 0 & -3 \end{bmatrix} x + \begin{bmatrix} -2 \\ 3 \\ 0 \end{bmatrix} u_1 + \begin{bmatrix} 1 \\ 0 \\ 1 \end{bmatrix} u_2$$
$$e_1 = \begin{bmatrix} 1 & 1 & 0 \end{bmatrix} x - (y_{ref})_1$$
$$e_2 = \begin{bmatrix} 0 & 0 & 1 \end{bmatrix} x - (y_{ref})_2.$$

   (a) Find the DC gain $G(0)$ of the system.
   (b) With $K = -G(0)^{-1}$ and the centralized controller (4.5)–(4.6) applied, find the range of $\varepsilon > 0$ for which we have closed-loop stability.
   (c) Carry out a decentralized design. Set $i_1 = 1$ and $i_2 = 2$; in the first step, find the range of ε_1 for closed-loop stability; in the second step, set ε_1 at the mid-point of the acceptable range, and then find the range of ε_2 for closed-loop stability.

5. Suppose that, given an ordering of the set $\{i \in \mathbf{N} : i = 1, \ldots \nu\}$, namely $i_1, .., i_\nu$, there exists a set of matrices $K_i \in \mathbf{R}^{m_i \times r_i}$ which satisfy conditions (i)...(ν) in Theorem 4.3. Prove that almost all such gain matrices will satisfy these conditions, in the sense that the set of matrices $\{K_1, \ldots, K_\nu\}$ which do not satisfy these conditions belongs to the zeros of a nonzero polynomial in the elements of these matrices.

References

1. G.W.M. Coppus, S.L. Shah, R.K. Wood, Robust multivariable control of a binary distillation column, in *IEE Proceedings Pt. D*, vol. 130, pp. 201–208 (1983)
2. E.J. Davison, Multivariable tuning regulators: the feedforward and robust control of a general servomechanism problem. IEEE Trans. Autom. Control **21**, 35–37 (1976)
3. E.J. Davison, Decentralized robust control of unknown systems using tuning regulators. IEEE Trans. Autom. Control **23**, 276–289 (1978)
4. E.J. Davison, P.A. Taylor, J.D. Wright, On the application of tuning regulators to control a commercial heat exchanger. IEEE Trans. Autom. Control **25**(3), 361 – 375 (1980)
5. E.J. Davison, A. Solomon, Partial decentralized temperature control of multi-zone buildings, in *22nd IEEE Conference on Decision and Control* (1983), pp. 10–16
6. E.J. Davison, N.K. Tripathi, Decentralized tuning regulators: an application to solve the load and frequency control problem for a large power system, in *18th IEEE Conference on Decision and Control* (1979), pp. 879–885
7. A.H. Etemadi, E.J. Davison, R. Iravani, A decentralized robust control strategy for Multi-DER microgrids - part I: fundamental concepts. IEEE Trans. Power Deliv. **27**(4), 1843–1853 (2012)
8. A.H. Etemadi, E.J. Davison, R. Iravani, A decentralized robust control strategy for Multi-DER microgrids-part II: performance evaluation. IEEE Trans. Power Deliv. **27**(4), 1854–1861 (2012)
9. G.H. Golub, C.F. Van Loan, *Matrix Computations* (The John Hopkins University Press, Baltimore, 1991)

10. H. Logemann, S. Townley, Low-gain control of uncertain regular linear systems. SIAM J. Control Optim. **35**(1), 78–116 (1997)
11. D.E. Miller, E.J. Davison, The self-tuning robust servomechanism problem. IEEE Trans. Autom. Control **34**, 511–523 (1989)
12. D.E. Miller, E.J. Davison, An adaptive tracking problem with a control input constraint. Automatica, 877–887 (1993)
13. J. Penttinen, H.N. Koivo, Multivariable tuning regulators for unknown systems. Automatica **16**(4), 393–398 (1980)

Chapter 5
Stability of Decentralized Systems Using Non-LTI Controllers

Up to this point in the book, we have focused our attention on the design of LTI controllers for the control of LTI decentralized systems. Now, we explore the possible benefits of more complicated nonlinear time-varying (NLTV) or linear time-varying (LTV) controllers; a special subclass of the latter are those which are the linear periodic controllers (LPCs), which will prove to be very useful. In this chapter, we focus on the stability problem; in Chap. 7, we focus on performance, especially near-optimal performance.

5.1 Introduction

As discussed in Chap. 2, and proven in [4, 5], the DFMs of a decentralized LTI system are immoveable using a decentralized LTI controller, which means that such a system can be stabilized using a decentralized LTI controller if and only if the system does not possess any unstable DFMs. However, it turns out that the situation is much different if we allow more complicated controllers.

An outline of this chapter is as follows:

- We first provide a simple example for which a linear time-varying controller can stabilize the plant while an LTI controller cannot.
- We then introduce the notion of a graph associated with a decentralized control system, and the accompanying notion of a quotient system. We define the quotient fixed modes[1] (QFMs), which are a subset of the system's DFMs. We provide some key results, including the fact that there exists a NLTV controller to provide closed-loop stability if and only if there are no unstable QFMs.
- We next provide a number of efficient ways to compute the QFMs.

---

[1]These are called "quotient decentralized fixed modes" in [13–16].

E. J. Davison et al., *Decentralized Control of Large-Scale Systems*,
https://doi.org/10.1007/978-1-4419-6014-6_5

- Last of all, we provide three systematic design methodologies for the design of LPCs which provide closed-loop stability in the event that the system has no unstable QFMs.

In this chapter, we consider the strictly proper plant

$$\left.\begin{aligned} \dot{x} &= Ax + \textstyle\sum_{i=1}^{\nu} B_i u_i, \quad x(t_0) = x_0 \\ y_i &= C_i x, \quad i = 1, \ldots, \nu \end{aligned}\right\} \tag{5.1}$$

with $x(t) \in \mathbf{R}^n$, $u_i(t) \in \mathbf{R}^{m_i}$ and $y_i(t) \in \mathbf{R}^{r_i}$ for $i = 1, \ldots, \nu$; we set $m = \sum_{i=1}^{\nu} m_i$ and $r = \sum_{i=1}^{\nu} r_i$; we represent this model by the data structure $(A; B_1, \ldots, B_p; C_1, \ldots, C_\nu; 0, \ldots, 0)$, with the zeros representing the D_{ij} terms which happen to be zero here. Associated with this model are

$$B := \begin{bmatrix} B_1 & \cdots & B_\nu \end{bmatrix},$$

$$y = \begin{bmatrix} y_1 \\ \vdots \\ y_\nu \end{bmatrix}, \quad u = \begin{bmatrix} u_1 \\ \vdots \\ u_\nu \end{bmatrix}, \quad C := \begin{bmatrix} C_1 \\ \vdots \\ C_\nu \end{bmatrix}.$$

In the decentralized context, we require u_i to depend solely on y_i, whether the controller is LTI, LTV or NLTV.

Before proceeding, we need some notation. We use the Euclidean norm for vectors and the corresponding induced norm for matrices, and denote the norm of a vector or matrix by $\|\cdot\|$. We say that $f : \mathbf{R}^+ \to \mathbf{R}^{n\times m}$ is of order T^j, and we write $f = \mathcal{O}(T^j)$, if there exist constants $c_1 > 0$ and $T_1 > 0$ so that

$$\|f(T)\| \le c_1 T^j, \quad T \in (0, T_1).$$

5.2 A Motivating Example

To illustrate the benefit of a non-LTI controller, let us consider a simple example:

$$\dot{x} = \begin{bmatrix} 1 & 0 & 0 \\ 0 & 0 & 1 \\ 0 & 2 & 1 \end{bmatrix} x + \begin{bmatrix} 0 \\ 0 \\ 1 \end{bmatrix} u_1 + \begin{bmatrix} 1 \\ 0 \\ 0 \end{bmatrix} u_2$$

$$y_1 = \begin{bmatrix} 1 & 0 & 0 \end{bmatrix} x$$

$$y_2 = \begin{bmatrix} 0 & -1 & 1 \end{bmatrix} x.$$

In transfer function form, this system corresponds to

$$\begin{bmatrix} Y_1(s) \\ Y_2(s) \end{bmatrix} = \begin{bmatrix} 0 & \frac{1}{s-1} \\ \frac{s-1}{(s+1)(s-2)} & 0 \end{bmatrix} \begin{bmatrix} U_1(s) \\ U_2(s) \end{bmatrix}.$$

It is easy to check that this has a decentralized mode at $s = 1$; indeed, with $k_i \in \mathbf{R}$,

$$K = \begin{bmatrix} k_1 & 0 \\ 0 & k_2 \end{bmatrix},$$

and

$$u = Ky,$$

we have that

$$det[sI - (A + BKC)] = (s - 1)[s^2 - s - 2 - k_1 k_2].$$

Now, let us consider a general controller of the form

$$\begin{bmatrix} u_1 \\ u_2 \end{bmatrix} = \begin{bmatrix} K_1 y_1 \\ K_2 y_2 \end{bmatrix},$$

with K_i possibly NLTV. To analyse closed-loop stability in the input/output setting, we inject noise d at the plant input and noise w at the plant output, and consider the closed-loop system given in Fig. 5.1. However, this can be rewritten as a single feedback loop as in Fig. 5.2.

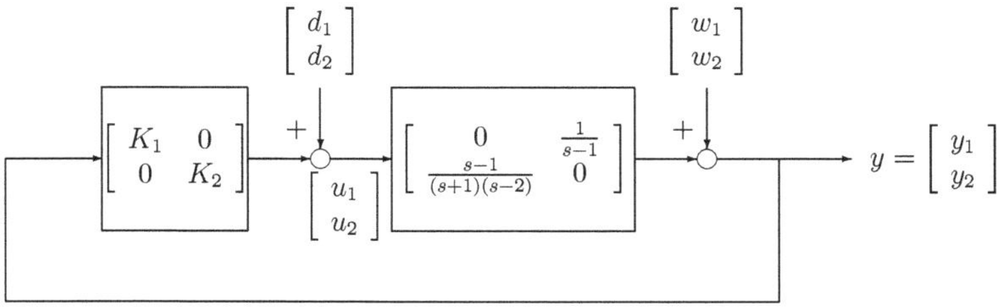

Fig. 5.1 The decentralized feedback control system

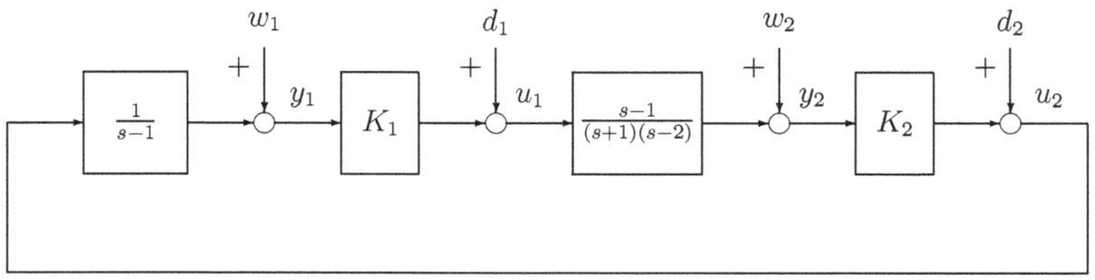

Fig. 5.2 The decentralized feedback control system as a single feedback loop

So if K_i are LTI with transfer functions of $\frac{f_1(s)}{g_1(s)}$ and $\frac{f_2(s)}{g_2(s)}$ accordingly, then it is easy to see that the "generalized characteristic polynomial" corresponding to the second feedback loop is

$$(s-1)g_1(s)(s+1)(s-2)g_2(s) + f_1(s)(s-1)f_2(s)$$

$$= (s-1)[(s+1)(s+2)g_1(s)g_2(s) + f_1(s)f_2(s)],$$

whose zeros determine the closed-loop poles; the presence of the zero at $s = 1$ guarantees that there is always a closed-loop pole at $s = 1$, regardless of the choice of the controller parameters.

Motivated by [17], suppose that we now consider a sampled-data controller: so we place a zero-order hold (ZOH) at the plant input and a sampler at the plant output, as in Fig. 5.3. With $T > 0$ the sampling period, if we adopt the natural notation of

$$\bar{x}[k] := x(kT), \ \bar{y}_i[k] := y_i(kT), \ \bar{d}_i[k] = d_i(kT), \ \bar{w}_i[k] = w_i(kT)$$

then we end up with the following state-space representation at the sample points:

$$\bar{x}[k+1] = e^{AT}\bar{x}[k] + \sum_{i=1}^{\nu}[\int_0^T e^{A\tau} B_i \, d\tau] \, \bar{u}_i[k]$$
$$\bar{y}_i[k] = C_i\bar{x}[k].$$

If we now apply the procedure to our example and apply the discrete-time controller

$$\bar{u}_i = \bar{K}_i \bar{y}_i,$$

then in closed loop, we end up with Fig. 5.4. There is no longer a pole-zero cancellation between the two plant components in the feedback loop for any $T > 0$: indeed, one plant component in the feedback loop has no zeros and the other has a zero at

$$\frac{4e^{2T} - 3e^{T} - e^{-T}}{e^{2T} - 4e^{-T} + 3}, \tag{5.2}$$

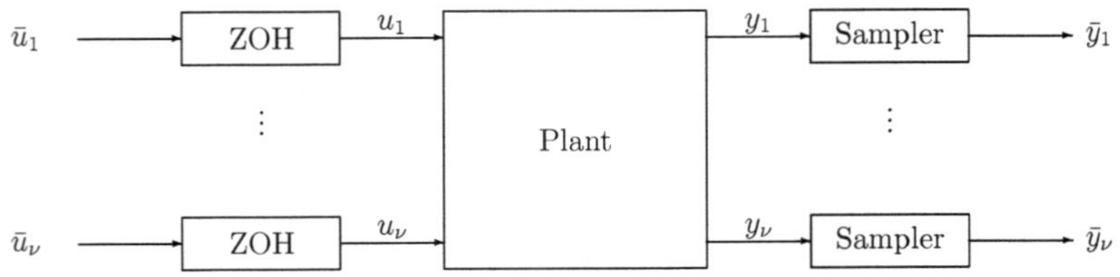

Fig. 5.3 The discretized system

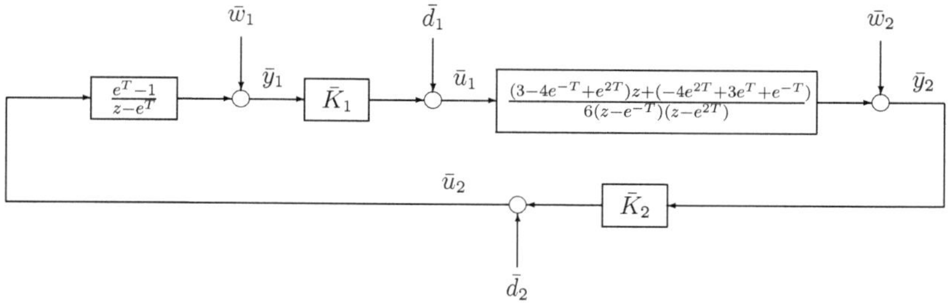

Fig. 5.4 The discretized closed-loop system as a single feedback loop

and it can be confirmed that this is never equal to the poles at e^{-T}, e^T and e^{2T}. (Of course, it is the pole at $z = e^T$ which is the one of concern.) Hence, there exist many possible discrete-time controllers which provide closed-loop stability; for example, we could set

$$\bar{K}_1[z] = 1, \ \ \bar{K}_2[z] = \frac{\bar{f}_2[z]}{\bar{g}_2[z]},$$

and then choose $\bar{f}_2[z]$ and $\bar{g}_2[z]$ so that the characteristic polynomial

$$6(z - e^T)(z - e^{-T})(z - e^{2T})\bar{g}_2[z] + [(3 - 4e^{-T} + e^{2T})z + (-4e^{2T} + 3e^T + e^{-T})]\bar{f}_2[z]$$

has all roots in the open unit disc. That being said, it turns out that the zero in (5.2) equals

$$e^T - \frac{1}{6}T^3 + \mathcal{O}(T^4)$$

(see Exercise 1), i.e. for small $T > 0$, there is a near pole-zero cancellation, which means that any LTI discrete-time controller designed will be fragile, e.g. the gain margin will be small, and the associated controller gains will be large. Of course, if $T > 0$ is large, say $T = 1$, then the zero and the problematic pole will be far apart: there is a pole at $z = 2.718$ and a zero at $z = 2.3587$; however, with such a large sampling period, the inter-sampling behaviour will be poor. Hence, while using a linear time-varying controller such as the sampled-data controller considered here may allow one to stabilize the plant, it often comes at the cost of poor performance. Last of all, notice that the overall controller is linear periodic (of period T) in the continuous-time domain, since it consists of the sampler, the zero-order hold and the DT controller.

Remark 1 The approach proposed here has limitations. For example, consider the system

$$\dot{x}(t) = \begin{bmatrix} 0 & 0 & 0 \\ 0 & 0 & 1 \\ 0 & 1 & 0 \end{bmatrix} x + \begin{bmatrix} 1 \\ 0 \\ 0 \end{bmatrix} u_1(t) + \begin{bmatrix} 0 \\ 0 \\ 1 \end{bmatrix} u_2(t)$$
$$y_1(t) = \begin{bmatrix} 0 & 0 & 1 \end{bmatrix} x(t)$$
$$y_2(t) = \begin{bmatrix} 1 & 0 & 0 \end{bmatrix} x(t)$$

with a transfer function of

$$\begin{bmatrix} 0 & \frac{s}{(s+1)(s-1)} \\ \frac{1}{s} & 0 \end{bmatrix}.$$

Using the graphical argument from above, it is clear that this system has a DFM at $s = 0$. It turns out that the discretized version of this system has a DFM at $z = 1$ for every $T > 0$. However, we shall soon see that this plant can be stabilized using a more sophisticated sampled-data controller.

5.3 The System Graph, the Quotient System and QFMs

Graph theory plays a key role in understanding certain aspects of decentralized systems. Following the approach of [3], we can create a directed graph of the plant (5.1) in the following way: there are ν nodes, with the ith one representing the ith channel, and there is a directed arc from node i to node j if and only if $C_j(sI - A)^{-1}B_i \neq 0$. Associated with this graph is its **topology matrix** $T \in \mathbf{R}^{p\times p}$ defined by

$$T_{ji} = \begin{cases} 0 & \text{if } C_j(sI - A)^{-1}B_i = 0 \\ 1 & \text{otherwise;} \end{cases}$$

this matrix is a concise way to represent the information needed to construct the graph of the system. Some directed graphs have very rich structure:

- It is said to be **complete** if there is a direct link, in both directions, between every two nodes, and
- It is said to be **strongly connected** if there is a path from every node to every other node along the arcs.

These concepts can be translated to a property of the topology matrix:

- The directed graph is complete if and only if every element of the topology matrix is one, and
- It is well-known that the directed graph is strongly connected if and only if every element of the matrix
$$T + T^2 + \cdots + T^{\nu}$$
is nonzero.

It turns out that a decentralized system with a strongly connected graph can always be stabilized by a linear time-varying controller under very modest conditions.

Theorem 1 [2] *If the plant (5.1) is centrally controllable and centrally observable and the associated graph is strongly connected, then it can be stabilized by a linear time-varying controller.*

Example 1 Consider the key example of Sect. 5.2:

$$\dot{x} = \begin{bmatrix} 1 & 0 & 0 \\ 0 & 0 & 1 \\ 0 & 2 & 1 \end{bmatrix} x + \begin{bmatrix} 0 \\ 0 \\ 1 \end{bmatrix} u_1 + \begin{bmatrix} 1 \\ 0 \\ 0 \end{bmatrix} u_2$$

$$y_1 = \begin{bmatrix} 1 & 0 & 0 \end{bmatrix} x$$

$$y_2 = \begin{bmatrix} 0 & -1 & 1 \end{bmatrix} x,$$

which is centrally controllable and observable; the associated transfer function is

$$\begin{bmatrix} Y_1(s) \\ Y_2(s) \end{bmatrix} = \begin{bmatrix} 0 & \frac{1}{s-1} \\ \frac{s-1}{(s+1)(s-2)} & 0 \end{bmatrix} \begin{bmatrix} U_1(s) \\ U_2(s) \end{bmatrix},$$

so the associated graph is clearly strongly connected. The above theorem says that there exists a linear time-varying controller which can stabilize it. This is consistent with the analysis of the previous section, wherein a sampled-data stabilizing controller is designed.

Example 2 Suppose that the plant transfer function (with the input and output partitioned into $\nu = 5$ components) look like this[2]:

$$C(sI-A)^{-1}B = \begin{bmatrix} C_1(sI-A)^{-1}B_1 & \cdots & C_1(sI-A)^{-1}B_5 \\ \vdots & & \vdots \\ C_5(sI-A)^{-1}B_1 & \cdots & C_5(sI-A)^{-1}B_5 \end{bmatrix} = \begin{bmatrix} 0 & 0 & \star & 0 & 0 \\ \star & 0 & 0 & 0 & 0 \\ 0 & \star & 0 & 0 & 0 \\ 0 & 0 & \star & 0 & \star \\ 0 & 0 & 0 & \star & 0 \end{bmatrix},$$

with a $\star$ denoting a nonzero transfer function. Then the associated topology matrix is

[2]This example is similar to that examined in [7].

$$T = \begin{bmatrix} 0\,0\,1\,0\,0 \\ 1\,0\,0\,0\,0 \\ 0\,1\,0\,0\,0 \\ 0\,0\,1\,0\,1 \\ 0\,0\,0\,1\,0 \end{bmatrix},$$

and

$$T + T^2 + T^3 + T^4 + T^5 = \begin{bmatrix} 1\,2\,2\,0\,0 \\ 2\,1\,2\,0\,0 \\ 2\,2\,1\,0\,0 \\ 2\,3\,4\,2\,3 \\ 1\,2\,3\,3\,2 \end{bmatrix};$$

since the latter matrix has some zero elements, the associated graph for the system cannot be strongly connected. Indeed, the graph is given in Fig. 5.5. It is clear that this graph is not strongly connected.

It turns out that the directed graph associated with the plant can be used to determine which DFMs are truly fixed and those which are not. Following [7], we can use the approach of [3] to construct a **quotient system** associated with the plant; this is another decentralized system obtained by grouping certain classes of channels together. It turns out that the directed graph can always be partitioned uniquely into a number of strongly connected subsystems, which can then be ordered in a very structured way. To do this, the graph is partitioned into $\bar{\nu}(\leq \nu)$ strongly connected subsystems, with the ith subsystem containing k_i input/output channels; this means, of course, that $\nu = \sum_{i=1}^{\bar{\nu}} k_i$. By relabelling the nodes (and the associated input/output channels) as necessary, and then partitioning them to correspond with the strongly connected subsystems, we will obtain a topology matrix which is in upper block diagonal form:

$$\begin{bmatrix} T_{11} & T_{12} & \cdots & T_{1\bar{\nu}} \\ 0 & T_{22} & \cdots & T_{2\bar{\nu}} \\ \vdots & \ddots & \ddots & \vdots \\ 0 & \cdots & 0 & T_{\bar{\nu}\bar{\nu}} \end{bmatrix};$$

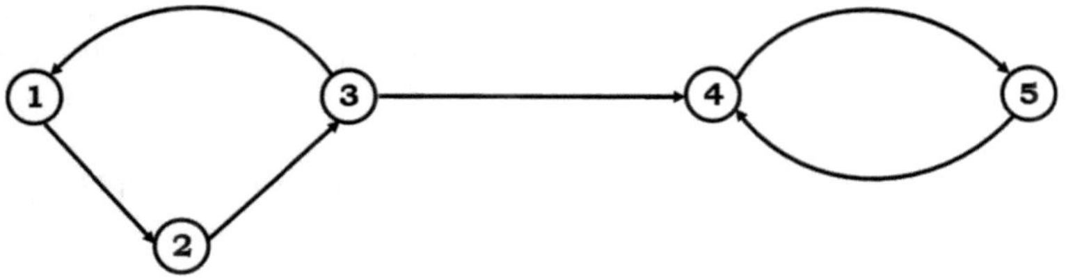

Fig. 5.5 The graph associated with Example 2

for every $i = 1, \ldots, \bar{\nu}$, we have $T_{ii} \in \mathbf{R}^{k_i \times k_i}$, and since it corresponds to a strongly connected graph in its own right, every element of

$$T_{ii} + T_{ii}^2 + \cdots + T_{ii}^{k_i}$$

will be nonzero. Indeed, since numbering of the input/outputs (or nodes on the graph) is arbitrary, for simplicity we may as well assume that the indices were already ordered in this fashion. At this point, we can construct a new decentralized control system commensurate with the above partitioning: we first repartition B, C, u and y as follows:

$$\bar{B}_1 := \begin{bmatrix} B_1 & \cdots & B_{k_1} \end{bmatrix}, \ \bar{B}_2 := \begin{bmatrix} B_{k_1+1} & \cdots & B_{k_1+k_2} \end{bmatrix}, \ldots, \bar{B}_{\bar{\nu}} := \begin{bmatrix} B_{\nu-k_{\bar{\nu}}+1} & \cdots & B_{\nu} \end{bmatrix}$$

$$\bar{C}_1 := \begin{bmatrix} C_1 \\ \vdots \\ C_{k_1} \end{bmatrix}, \ \bar{C}_2 := \begin{bmatrix} C_{k_1+1} \\ \vdots \\ C_{k_1+k_2} \end{bmatrix}, \ldots, \bar{C}_{\bar{\nu}} := \begin{bmatrix} C_{\nu-k_{\bar{\nu}}+1} \\ \vdots \\ C_{\nu} \end{bmatrix}$$

$$\bar{u}_1 := \begin{bmatrix} u_1 \\ \vdots \\ u_{k_1} \end{bmatrix}, \ \bar{u}_2 := \begin{bmatrix} u_{k_1+1} \\ \vdots \\ u_{k_1+k_2} \end{bmatrix}, \ldots, \bar{u}_{\bar{\nu}} := \begin{bmatrix} u_{\nu-k_{\bar{\nu}}+1} \\ \vdots \\ u_{\nu} \end{bmatrix},$$

$$\bar{y}_1 := \begin{bmatrix} y_1 \\ \vdots \\ y_{k_1} \end{bmatrix}, \ \bar{y}_2 := \begin{bmatrix} y_{k_1+1} \\ \vdots \\ y_{k_1+k_2} \end{bmatrix}, \ldots, \bar{y}_{\bar{\nu}} := \begin{bmatrix} y_{\nu-k_{\bar{\nu}}+1} \\ \vdots \\ y_{\nu} \end{bmatrix};$$

the **quotient system** is defined by

$$\left. \begin{aligned} \dot{x} &= Ax + \textstyle\sum_{i=1}^{\nu} \bar{B}_i \bar{u}_i \\ \bar{y}_i &= \bar{C}_i x, \quad i = 1, \ldots, \bar{\nu}. \end{aligned} \right\} \tag{5.3}$$

It turns out that it is possible to carry out a carefully constructed similarity transformation so that in the new coordinate system A, B and C are in upper block triangular form, with the dimensions of the blocks commensurate with the above partitioning; see [3] and Exercise 5 for more details.

Example 3 To illustrate the procedure of constructing a quotient system from a graphical point of view, let us return to Example 2. In this case, the quotient system is given by Fig. 5.6. So we see that the system can be partitioned into two strongly connected subsystems. In Fig. 5.7, we see the original system with a decentralized controller attached, and in Fig. 5.8, we see the quotient system with a (quotient) decentralized controller attached.

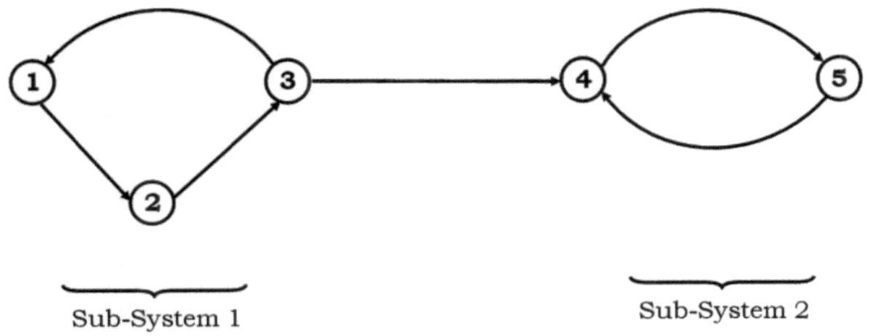

Fig. 5.6 The graph of the system partitioned into strongly connected subsystems

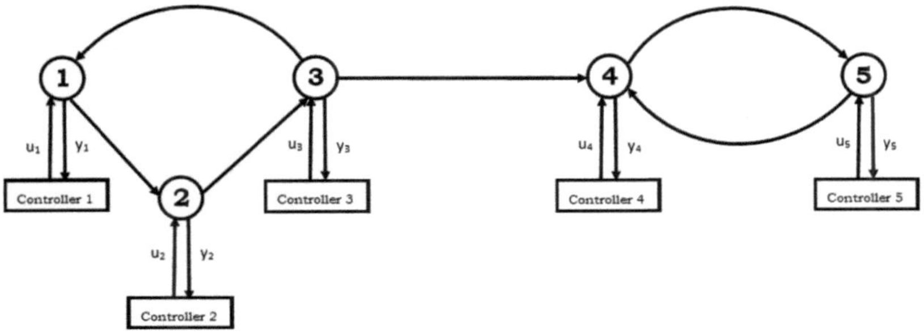

Fig. 5.7 The decentralized system with a controller applied

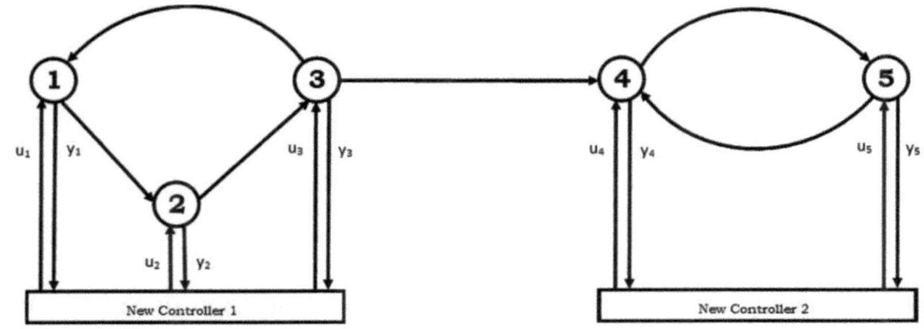

Fig. 5.8 The quotient system with a controller applied

The DFMs of the quotient system are labelled the **quotient fixed modes** (QFMs) of the original plant; clearly, they are a subset of its DFMs.

Example 4 To illustrate the procedure of constructing a quotient system using the matrix representation of the plant, consider a system with three input and output pairs, with a (centrally) controllable and observable state-space representation as follows:

$$\dot{x} = \begin{bmatrix} 0 & 1 & 0 & 0 & 0 & 0 & 0 \\ -2 & 3 & 0 & 0 & 0 & 0 & 0 \\ 0 & 0 & 0 & 1 & 0 & 0 & 0 \\ 0 & 0 & -1 & -2 & 0 & 0 & 0 \\ 0 & 0 & 0 & 0 & 3 & 0 & 0 \\ 0 & 0 & 0 & 0 & 0 & 0 & 1 \\ 0 & 0 & 0 & 0 & 0 & -4 & -4 \end{bmatrix} x + \begin{bmatrix} 0 \\ 0 \\ 0 \\ 1 \\ 0 \\ 0 \\ 0 \end{bmatrix} u_1 + \begin{bmatrix} 0 \\ 1 \\ 0 \\ 0 \\ 0 \\ 0 \\ 0 \end{bmatrix} u_2 + \begin{bmatrix} 0 \\ 0 \\ 0 \\ 0 \\ 1 \\ 0 \\ 1 \end{bmatrix} u_3$$

$$y_1 = \begin{bmatrix} 0 & 1 & 0 & 0 & 1 & 0 & 0 \end{bmatrix} x$$
$$y_2 = \begin{bmatrix} 0 & 0 & -1 & 1 & 0 & 0 & 0 \end{bmatrix} x$$
$$y_3 = \begin{bmatrix} 0 & 0 & 0 & 0 & 0 & -3 & 1 \end{bmatrix} x.$$

This has a transfer function of

$$\begin{bmatrix} Y_1(s) \\ Y_2(s) \\ Y_3(s) \end{bmatrix} = \begin{bmatrix} 0 & \frac{s}{(s-1)(s-2)} & \frac{1}{s-3} \\ \frac{s-1}{(s+1)^2} & 0 & 0 \\ 0 & 0 & \frac{s-3}{(s+1)^2} \end{bmatrix} \begin{bmatrix} U_1(s) \\ U_2(s) \\ U_3(s) \end{bmatrix}.$$

The graph of this system is in Fig. 5.9 it is clear that it is not strongly connected. Indeed, it has two strongly connected subsets: $\{1, 2\}$ and $\{3\}$. Hence, the corresponding quotient system is given by

$$\dot{x} = \begin{bmatrix} 0 & 1 & 0 & 0 & 0 & 0 & 0 \\ -2 & 3 & 0 & 0 & 0 & 0 & 0 \\ 0 & 0 & 0 & 1 & 0 & 0 & 0 \\ 0 & 0 & -1 & -2 & 0 & 0 & 0 \\ 0 & 0 & 0 & 0 & 3 & 0 & 0 \\ 0 & 0 & 0 & 0 & 0 & 0 & 1 \\ 0 & 0 & 0 & 0 & 0 & -4 & -4 \end{bmatrix} x + \begin{bmatrix} 0 & 0 \\ 0 & 1 \\ 0 & 0 \\ 1 & 0 \\ 0 & 0 \\ 0 & 0 \\ 0 & 0 \end{bmatrix} \underbrace{\begin{bmatrix} u_1 \\ u_2 \end{bmatrix}}_{=:\bar{u}_1} + \begin{bmatrix} 0 \\ 0 \\ 0 \\ 0 \\ 1 \\ 0 \\ 1 \end{bmatrix} \underbrace{u_3}_{=:\bar{u}_2}$$

$$\bar{y}_1 = \begin{bmatrix} y_1 \\ y_2 \end{bmatrix} = \begin{bmatrix} 0 & 1 & 0 & 0 & 1 & 0 & 0 \\ 0 & 0 & -1 & 1 & 0 & 0 & 0 \end{bmatrix} x$$
$$\bar{y}_2 := y_3 = \begin{bmatrix} 0 & 0 & 0 & 0 & 0 & -3 & 1 \end{bmatrix} x.$$

It is easy to verify numerically that the original system has DFMs at $s = 1, 3$, while the quotient system has a DFM at $s = 3$; this means that the original system has a QFM at $s = 3$.

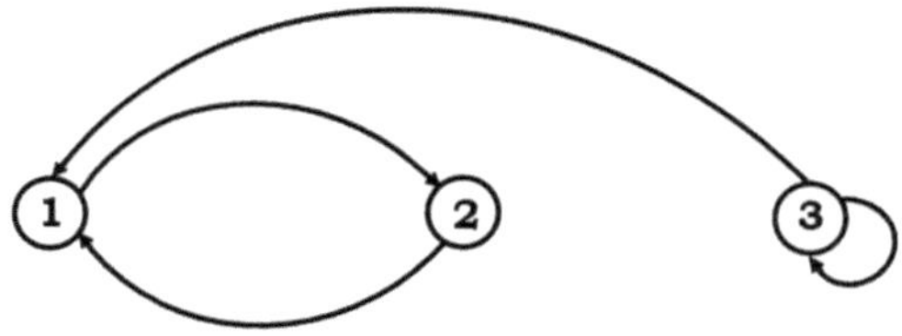

Fig. 5.9 The graph for Example 4

Under some stringent conditions, it turns out that the system has no QFMs.

Theorem 2 *If (A, B) is controllable, (C, A) is observable, and the graph associated with the system is strongly connected, then the system has no QFMs.*

Proof If the graph is strongly connected, then the quotient system is simply the centralized counterpart. Since it is controllable and observable by hypothesis, the centralized system clearly has no fixed modes, so the original system has no QFMs. □

It turns out that the existence of QFMs plays an important role, in that they provide a clear delineation between when a decentralized system can and cannot be stabilized by a sufficiently complicated controller.

Theorem 3 *[7] There exists a NLTV controller to stabilize the decentralized plant if and only if it has no unstable QFMs.*

Idea of the proof: The proof of sufficiency provided in [7] has three steps:

(i) As discussed above, use graph theory to form the quotient system.
(ii) Motivated by [2], for each strongly connected subsystem, a decentralized time-varying output feedback gain is designed which makes that subsystem controllable and observable via a single channel; while [2] only considers the two-channel case in detail, in principle, it can be applied iteratively if there are more than two channels.
(iii) Using [8], for each strongly connected subsystem, one can design a linear time-varying observer-based stabilizing controller to be applied on the channel from which the system is controllable and observable.

□

While the proof provided above is constructive, the proposed controller is not particularly easy to design. At this point, we would like to pursue two different avenues:

(i) First of all, checking the system's QFMs requires two steps: analyse its graph and partition it into strongly connected components, followed by applying standard tools of decentralized control to ascertain the DFMs of the quotient system; the problem is that the first step can be numerically fragile. It would be desirable to have a more direct way to compute the QFMs. We do so in the next section.
(ii) Second of all, if there are no unstable QFMs, then the controller proven to exist by [7] is not easy to construct. It would be desirable to have a straightforward and algorithmic design procedure. We provide several such methods.

5.4 A Non-graphical Way to Compute QFMs

In this section, we provide three different ways to compute QFMs in an efficient manner; the first two require an additional assumption, but the third way does not. To this end, given the decentralized plant (5.1), represented by $(A;\ B_1, \dots, B_p; C_1, \dots, C_p;\ 0, \dots, 0)$, we now construct three related decentralized systems, which were introduced in [13, 14].

Model Based on Controllability: With

$$B_i^{con} := \left[\, B_i \ AB_i \ \cdots \ A^{n-1}B_i \,\right] \in \mathbf{R}^{n\times nm_i}, \quad i = 1, \dots, \nu,$$

$$B^{con} := \left[\, B_1^{con} \ \cdots \ B_\nu^{con} \,\right],$$

we construct the first model by replacing each B_i by the corresponding controllability matrix, yielding

$$S_{con} = (A;\ B_1^{con}, \dots, B_\nu^{con};\ C_1, \dots, C_\nu;\ 0, \dots, 0).$$

Model Based on Observability: With

$$C_i^{obs} := \begin{bmatrix} C_i \\ C_i A \\ \vdots \\ C_i A^{n-1} \end{bmatrix} \in \mathbf{R}^{nr_i\times n}, \quad i = 1, \dots, \nu, \quad C^{obs} := \begin{bmatrix} C_1^{obs} \\ \vdots \\ C_\nu^{obs} \end{bmatrix},$$

we construct the second model by replacing each C_i by the corresponding observability matrix, yielding

$$S_{obs} = (A;\ B_1, \dots, B_\nu;\ C_1^{obs}, \dots, C_\nu^{obs};\ 0, \dots, 0).$$

Model Based on Markov Parameters: In this model, we replace the D_{ij}'s (which equal zero) by a matrix containing the first n Markov parameters, and we pad the B_i's and C_i's with zeros accordingly:

$$C_i^{mar} := \begin{bmatrix} C_i \\ 0 \\ \vdots \\ 0 \end{bmatrix} \in \mathbf{R}^{2nr_i\times n}, \quad D_{ij}^{mar} = \begin{bmatrix} 0 & & \\ C_iB_j & \ddots & \\ \vdots & \ddots & 0 \\ C_iA^{n-1}B_j & \cdots & C_iB_j \\ & \ddots & \vdots \\ & & C_iA^{n-1}B_j \end{bmatrix} \in \mathbf{R}^{2nr_i\times nm_i},$$

$$B_i^{mar} := \begin{bmatrix} B_i & 0 & \cdots & 0 \end{bmatrix} \in \mathbf{R}^{n \times nm_i}, \quad B^{mar} := \begin{bmatrix} B_1^{mar} & \cdots & B_\nu^{mar} \end{bmatrix}, \quad C^{mar} := \begin{bmatrix} C_1^{mar} \\ \vdots \\ C_\nu^{mar} \end{bmatrix},$$

and define S_{mar} by

$$S_{mar} = (A; B_1^{mar}, \ldots, B_\nu^{mar}; C_1^{mar}, \ldots, C_\nu^{mar}; D_{11}^{mar}, \ldots, D_{1\nu}^{mar}, \ldots D_{\nu 1}^{mar}, \ldots, D_{\nu\nu}^{mar}).$$

Theorem 4 *[14]*

(i) The QFMs of the original system (5.1) are the DFMs of S_{mar}.
(ii) If there is only one Jordan block for every eigenvalue of A, then the QFMs of the original system (5.1) are the DFMs of S_{con}.
(iii) If there is only one Jordan block for every eigenvalue of A, then the QFMs of the original system (5.1) are the DFMs of S_{obs}.

Remark 2 While checking the size of the Jordan blocks is numerically fragile in general, finding the eigenvalues of A is numerically robust: hence, we can apply parts (ii) or (iii) if A has distinct eigenvalues, and otherwise, we can simply apply part (i). In any event, we can find the QFMs simply by finding the DFMs of a related system, so we remove a step from the two-step procedure for finding QFMs discussed in the previous section.

Example 5 Consider the example of Remark 1: we have two channels and

$$A := \begin{bmatrix} 0 & 0 & 0 \\ 0 & 0 & 1 \\ 0 & 1 & 0 \end{bmatrix}, \quad B_1 := \begin{bmatrix} 1 \\ 0 \\ 0 \end{bmatrix}, \quad B_2 := \begin{bmatrix} 0 \\ 0 \\ 1 \end{bmatrix},$$

$$C_1 := \begin{bmatrix} 0 & 0 & 1 \end{bmatrix}, \quad C_2 := \begin{bmatrix} 1 & 0 & 0 \end{bmatrix};$$

it is observed there that the system has an unstable DFM at $s = 0$, and the discretized counterpart has an unstable DFM at $z = 1$. While Theorem 2 and the fact that the graph is strongly connected mean that the system has no QFMs, let us now apply Theorem 4 to derive the same result without resorting to the use of a graph. Since there are three eigenvalues at $\{0, -1, 1\}$ and $n = 3$, every eigenvalue of A has only one Jordan block, so we can apply Theorem 3 (ii). To this end, we form a state-space model of S_{con}:

$$\begin{aligned}
\dot{x}^{con} &= \begin{bmatrix} 0\,0\,0 \\ 0\,0\,1 \\ 0\,1\,0 \end{bmatrix} x^{con} + \begin{bmatrix} 1\,0\,0 \\ 0\,0\,0 \\ 0\,0\,0 \end{bmatrix} u_1^{con} + \begin{bmatrix} 0\,0\,0 \\ 0\,1\,0 \\ 1\,0\,1 \end{bmatrix} u_2^{con} \\
y_1^{con} &= \begin{bmatrix} 0\,0\,1 \end{bmatrix} x^{con} \\
y_2^{con} &= \begin{bmatrix} 1\,0\,0 \end{bmatrix} x^{con}.
\end{aligned}$$

To find its DFMs, we simply apply some output feedback:

$$u_1^{con} = \begin{bmatrix} k_1 \\ k_2 \\ k_3 \end{bmatrix} y_1^{con} =: K_1^{con} y_1^{con}$$

$$u_2^{con} = \begin{bmatrix} k_4 \\ k_5 \\ k_6 \end{bmatrix} y_2^{con} =: K_2^{con} y_2^{con},$$

so

$$\begin{aligned}
det(sI - A - B_1^{con} K_1^{con} C_1^{con} - B_2^{con} K_2^{con} C_2^{con}) &= \begin{bmatrix} s & 0 & -k_1 \\ -k_5 & s & -1 \\ -k_4 - k_6 & -1 & s \end{bmatrix} \\
&= s^3 - k_1 k_5 - k_1(k_4 + k_6)s - s \\
&= s^3 - (1 + k_1 k_4 + k_1 k_6)s - k_1 k_5.
\end{aligned}$$

It is easy to see that all three roots vary as a function of the free parameters, so S_{con} has no DFMs, which means that the original system has no QFMs. Hence, while there does not exist an LTI controller which will stabilize the original system, and not even a sampled-data controller with a discrete-time LTI component, by Theorem 3 there does indeed exist a NLTV one which will do so. Indeed, we shall see in the next section that there exists a linear periodic sampled-data controller which can do so.

Example 6 Consider the decentralized control system

$$\begin{aligned}
\dot{x} &= \begin{bmatrix} 1\,0\,0\,0 \\ 0\,1\,0\,0 \\ 0\,0\,2\,0 \\ 0\,0\,0\,3 \end{bmatrix} x + \begin{bmatrix} 1 \\ 0 \\ 0 \\ 0 \end{bmatrix} u_1 + \begin{bmatrix} 0 \\ 1 \\ 1 \\ 1 \end{bmatrix} u_2, \\
y_1 &= \begin{bmatrix} 1\,0\,0\,1 \end{bmatrix} x \\
y_2 &= \begin{bmatrix} 0\;2\;-1\;0 \end{bmatrix} x.
\end{aligned}$$

It is easy to verify numerically that there is a DFM as $s = 3$. To compute the QFMs, we apply Theorem 4; since there are repeated eigenvalues, we will use part (i). To do so, we first form S_{mar}:

$$\dot{x}^{mar} = \begin{bmatrix} 1&0&0&0 \\ 0&1&0&0 \\ 0&0&2&0 \\ 0&0&0&3 \end{bmatrix} x^{mar} + \begin{bmatrix} 1&0&0&0 \\ 0&0&0&0 \\ 0&0&0&0 \\ 0&0&0&0 \end{bmatrix} u_1^{mar} + \begin{bmatrix} 0&0&0&0 \\ 1&0&0&0 \\ 1&0&0&0 \\ 1&0&0&0 \end{bmatrix} u_2^{mar}$$

$$y_1^{mar} = \begin{bmatrix} 1&0&0&1 \\ 0&0&0&0 \\ 0&0&0&0 \\ 0&0&0&0 \\ 0&0&0&0 \\ 0&0&0&0 \\ 0&0&0&0 \\ 0&0&0&0 \end{bmatrix} x^{mar} + \begin{bmatrix} 0&0&0&0 \\ 1&0&0&0 \\ 1&1&0&0 \\ 1&1&1&0 \\ 1&1&1&1 \\ 0&1&1&1 \\ 0&0&1&1 \\ 0&0&0&1 \end{bmatrix} u_1^{mar} + \begin{bmatrix} 0&0&0&0 \\ 1&0&0&0 \\ 3&1&0&0 \\ 9&3&1&0 \\ 27&9&3&1 \\ 0&27&9&3 \\ 0&0&27&9 \\ 0&0&0&27 \end{bmatrix} u_2^{mar},$$

$$y_2^{mar} = \begin{bmatrix} 0&2&-1&0 \\ 0&0&0&0 \\ 0&0&0&0 \\ 0&0&0&0 \\ 0&0&0&0 \\ 0&0&0&0 \\ 0&0&0&0 \\ 0&0&0&0 \end{bmatrix} x^{mar} + \begin{bmatrix} 0&0&0&0 \\ 0&0&0&0 \\ 0&0&0&0 \\ 0&0&0&0 \\ 0&0&0&0 \\ 0&0&0&0 \\ 0&0&0&0 \\ 0&0&0&0 \end{bmatrix} u_1^{mar} + \begin{bmatrix} 0&0&0&0 \\ 1&0&0&0 \\ 0&1&0&0 \\ -2&0&1&0 \\ -6&-2&0&1 \\ 0&-6&-2&0 \\ 0&0&-6&-2 \\ 0&0&0&-6 \end{bmatrix} u_2^{mar}.$$

It is easy to verify numerically that this system has a DFM at $s = 3$, which means that the original system has a QFM at $s = 3$.

Remark 3 Another advantage of the algebraic approach proposed here over the graphical approach of Gong and Aldeen [7] is that the former can warn of a nearby QFM; details can be found in [14].

Now, we turn our attention to controller synthesis. Suppose that the plant has unstable DFMs but all QFMs are stable; the goal is to design a stabilizing controller. While Gong and Aldeen [7] provide a design procedure for a stabilizing nonlinear time-varying controller, it is quite involved and provides only a high-level design procedure without the low-level details. At this point, we present three different controller design algorithms which yield a **linear periodic sampled-data controller**:

(i) In the next section, we impose the assumption that for every eigenvalue of A there is exactly one Jordan block, so that we can apply Theorem 4 (ii). We analyse S_{con} and use it to motivate the design of a stabilizing controller for the nominal system using a generalized hold.
(ii) In the following section, we also impose the assumption that for every eigenvalue of A there is exactly one Jordan block, so that we can apply Theorem 4 (iii). We analyse S_{obs} and use it to motivate the design of a stabilizing controller for the nominal system using fast sampling and estimation.

(iii) In the last section, we consider the general case where some eigenvalues of A have more than one Jordan block, in which case we turn to a more complicated controller design procedure; because of the complexity of the approach, we provide the high-level intuition and refer the reader to [16] for details.

5.5 A Controller Design Based on S_{con} Using Generalized Holds

In this section, we assume that the system has no unstable QFMs and that for every eigenvalue of A there is exactly one Jordan block, so that we can apply Theorem 4 (ii). To this end, let us first form a state-space representation of S_{con}:

$$\begin{aligned} \dot{x}^{con} &= Ax^{con} + \sum_{i=1}^{\nu} B_i^{con} u_i^{con} \\ y_i^{con} &= C_i x^{con}; \end{aligned}$$

in condensed form, this becomes

$$\begin{aligned} \dot{x}^{con} &= Ax^{con} + B^{con} u^{con} \\ y^{con} &= Cx^{con}. \end{aligned}$$

It follows that S_{con} has no unstable DFMs; this means that there exists an LTI decentralized stabilizing controller K^{con} for this system:

$$\dot{z}_i^{con} = F_i z_i^{con} + G_i y_i^{con} \tag{5.4}$$

$$u_i^{con} = H_i z_i^{con} + J_i y_i^{con}; \tag{5.5}$$

here, $z_i^{con}(t) \in \mathbf{R}^{\ell_i}$, and we define $\ell := \ell_1 + \cdots + \ell_p$. Now, let us obtain a concise description of the decentralized stabilizing controller for S_{con}: with

$$F := \begin{bmatrix} F_1 & & \\ & \ddots & \\ & & F_p \end{bmatrix}, \quad G := \begin{bmatrix} G_1 & & \\ & \ddots & \\ & & G_p \end{bmatrix}, \quad H := \begin{bmatrix} H_1 & & \\ & \ddots & \\ & & H_p \end{bmatrix},$$

$$J := \begin{bmatrix} J_1 & & \\ & \ddots & \\ & & J_p \end{bmatrix}, \quad z^{con} := \begin{bmatrix} z_1^{con} \\ \vdots \\ z_p^{con} \end{bmatrix},$$

it follows that

$$\begin{aligned} \dot{z}^{con} &= Fz^{con} + Gy^{con} \\ u^{con} &= Hz^{con} + Jy^{con}. \end{aligned}$$

If we apply this to S_{con}, then the fictitious closed-loop system becomes

$$\begin{bmatrix} \dot{x}^{con} \\ \dot{z}^{con} \end{bmatrix} = \begin{bmatrix} A + B^{con} JC & B^{con} H \\ GC & F \end{bmatrix} \begin{bmatrix} x^{con} \\ z^{con} \end{bmatrix} =: A_{cl}^{con} \begin{bmatrix} x^{con} \\ z^{con} \end{bmatrix};$$

clearly, A_{cl}^{con} is stable.

The goal now is to use the above to design a sampled-data stabilizing controller or period T for the actual system (5.1); the underlying concept used here is the notion of a generalized hold [9]. To this end, suppose that we use the following discretized version of the above decentralized stabilizing controller K^{con} driven by the plant output:

$$\begin{aligned} z_i((k+1)T) &= e^{F_i T} z_i(kT) + [\int_0^T e^{F_i \tau} G_i d\tau] y_i(kT) \\ u_i(kT) &= H_i z_i(kT) + J_i y_i(kT); \end{aligned}$$

defining z in the natural way, this can be written in condensed form as

$$z((k+1)T) = e^{FT} z(kT) + [\int_0^T e^{F\tau} G d\tau] y(kT) \quad (5.6)$$

$$u(kT) = Hz(kT) + Jy(kT). \quad (5.7)$$

We then connect the output of this controller to the plant using a so-called generalized hold: with $L_i(t)$ of period T, set

$$u_i(t) = L_i(t) u_i(kT), \ t \in [kT, (k+1)T). \quad (5.8)$$

If we solve the plant equation at time instants which are integer multiples of T, then we have

$$x((k+1)T) = e^{AT} x(kT) + \sum_{i=1}^{\nu} [\int_0^T e^{A(T-\tau)} B_i L_i(\tau) d\tau] u_i(kT).$$

Using the notion of reachability in the study of linear systems, it is easy to see that we can always choose $L_i(t)$ so that

$$\int_0^T e^{A(T-\tau)} B_i L_i(\tau) d\tau$$

equals any matrix of appropriate dimension in the image of the corresponding controllability matrix

$$\begin{bmatrix} B_i & AB_i & \cdots & A^{n-1} B_i \end{bmatrix}$$

that we desire; here, we will choose a bounded $L_i(t)$ so that

$$\int_0^T e^{A(T-\tau)} B_i L_i(\tau) d\tau = T B_i^{con}; \tag{5.9}$$

this is not unique, so let us fix one such choice, and without loss of generality, we may as well assume that it is periodic of period T. With this choice of L_i, we have

$$x((k+1)T) = e^{AT} x(kT) + T \sum_{i=1}^{\nu} B_i^{con} u_i(kT) = e^{AT} x(kT) + T B^{con} u(kT).$$

If we now examine the closed-loop system, we see that

$$\begin{bmatrix} x((k+1)T) \\ z((k+1)T) \end{bmatrix} = \underbrace{\begin{bmatrix} e^{AT} + T B^{con} J C & T B^{con} H \\ (\int_0^T e^{F\tau} G d\tau) C & e^{FT} \end{bmatrix}}_{=: \hat{A}_{cl}^{con}(T)} \begin{bmatrix} x(kT) \\ z(kT) \end{bmatrix}. \tag{5.10}$$

So the controller consists of the discrete-time compensator (5.6)–(5.7) together with the generalized hold (5.8) satisfying (5.9). Given that the controller state is in discrete time while the plant state is in continuous time, we define the sampled-data state by

$$x_{sd}(t) := \begin{bmatrix} x(t) \\ z(kT) \end{bmatrix}, \quad t \in [kT, (k+1)T).$$

Theorem 5 *Suppose that the controller given by (5.6)–(5.7) together with the generalized hold (5.8) satisfying (5.9) is applied to the plant (5.1) at time $t_0 = 0$. Then there exists a constant $\bar{T} > 0$ so that for every $T \in (0, \bar{T})$ there exist constants $c > 0$ and $\lambda < 0$ so that for every $x_0 \in \mathbf{R}^n$ and $z(0) \in \mathbf{R}^\ell$, we have*

$$\|x_{sd}(t)\| \leq c e^{\lambda t} \|x_{sd}(0)\|, \quad t \geq 0.$$

Proof Let $x_0 \in \mathbf{R}^n$, $z(0) \in \mathbf{R}^\ell$, and $T > 0$ be arbitrary. It follows from (5.10) that the closed-loop behaviour at the sample points is governed by $\hat{A}_{cl}^{con}(T)$. But we see that

$$\begin{aligned} \hat{A}_{cl}^{con}(T) &= \begin{bmatrix} e^{AT} + T B^{con} J C & T B^{con} H \\ (\int_0^T e^{F\tau} G d\tau) C & e^{FT} \end{bmatrix} \\ &= \begin{bmatrix} I + AT + T B^{con} J C & T B^{con} H \\ T G C & I + FT \end{bmatrix} + \mathcal{O}(T^2) \\ &= I + A_{cl}^{con} T + \mathcal{O}(T^2). \end{aligned}$$

Since A_{cl}^{con} is stable, there exists a positive definite and symmetric square matrix P which satisfies

$$[A_{cl}^{con}]^T P + P A_{cl}^{con} + I = 0;$$

define the Lyapunov function

$$V(x_{sd}(kT)) := x_{sd}(kT)^T P x_{sd}(kT).$$

It is easy to see that

$$\begin{aligned} V(x_{sd}((k+1)T)) &= V(x_{sd}(kT)) - T\|x_{sd}(kT)\|^2 + \mathcal{O}(T^2)\|x_{sd}(kT)\|^2 \\ &\leq V(x_{sd}(kT)) - T\frac{1}{\|P\|}V(x_{sd}(kT)) + \mathcal{O}(T^2)V(x_{sd}(kT)); \end{aligned}$$

for small T, say $T \in (0, \bar{T})$, we have

$$V(x_{sd}((k+1)T)) \leq [1 - T\frac{1}{2\|P\|}]V(x_{sd}(kT)),$$

which means that

$$V(x_{sd}(kT)) \leq e^{-kT\frac{1}{2\|P\|}} V(x_{sd}(0)), \ k \geq 0.$$

With $\lambda_{min}(P)$ denoting the smallest eigenvalue of P, it follows that

$$\lambda_{min}(P)\|x_{sd}(kT)\|^2 \leq V(x_{sd}(kT)) \leq e^{-kT\frac{1}{2\|P\|}} \|P\| \times \|x_{sd}(0)\|^2, \ k \geq 0;$$

so if we set

$$c_1 := [\frac{\|P\|}{\lambda_{min}(P)}]^{1/2}$$

and $\lambda := -\frac{1}{4\|P\|}$, we see that for $T \in (0, \bar{T})$, we have

$$\|x_{sd}(kT)\| \leq c_1 e^{\lambda kT}\|x(0)\|, \ k \geq 0.$$

Since $L_i(t)$ is bounded, we can easily prove that nothing untoward happens to $x(t)$ between sample points, so we conclude that there for every $T \in (0, \bar{T})$, there exists a constant c_2 so that

$$\|x_{sd}(t)\| \leq c_2 e^{\lambda t}\|x(0)\|, \ t \geq 0.$$

□

Example 7 Here, we illustrate the design procedure proposed here. Consider the plant

$$\dot{x} = \begin{bmatrix} 0 & 0 & 0 \\ 0 & 0 & 1 \\ 0 & 4 & -3 \end{bmatrix} x + \begin{bmatrix} 1 \\ 0 \\ 0 \end{bmatrix} u_1 + \begin{bmatrix} 0 \\ 0 \\ 1 \end{bmatrix} u_2$$
$$y_1 = \begin{bmatrix} 0 & 0 & 1 \end{bmatrix} x$$
$$y_2 = \begin{bmatrix} 1 & 0 & 0 \end{bmatrix} x,$$

with a corresponding transfer function of

$$\begin{bmatrix} Y_1(s) \\ Y_2(s) \end{bmatrix} = \begin{bmatrix} 0 & \frac{s}{(s-1)(s+4)} \\ \frac{1}{s} & 0 \end{bmatrix} \begin{bmatrix} U_1(s) \\ U_2(s) \end{bmatrix}.$$

From the transfer function, we see that there is a DFM at $s = 0$; however, since the system is centrally controllable and centrally observable, and the associated graph is clearly strongly connected, it follows from Theorem 2 that it has no QFMs; from Theorem 1, we see that there exists a linear time-varying stabilizing controller. It can be easily checked that the discretized version of this plant has an unstable DFM at $z = 1$, so we cannot simply use a simple sampled-data controller considered in Sect. 5.2. However, it is easy to check that A has distinct eigenvalues, so Theorem 4(ii) says that the associated system S_{con} has no DFMs, which means that the procedure adopted in this section can be employed to design a linear periodic stabilizing controller.

To this end, we first form the associated model S_{con}: we have

$$\dot{x}^{con} = \begin{bmatrix} 0 & 0 & 0 \\ 0 & 0 & 1 \\ 0 & 4 & -3 \end{bmatrix} x^{con} + \begin{bmatrix} 1 & 0 & 0 \\ 0 & 0 & 0 \\ 0 & 0 & 0 \end{bmatrix} u_1^{con} + \begin{bmatrix} 0 & 0 & 0 \\ 0 & 1 & -3 \\ 1 & -3 & 13 \end{bmatrix} u_2^{con}$$
$$y_1^{con} = \begin{bmatrix} 0 & 0 & 1 \end{bmatrix} x$$
$$y_2^{con} = \begin{bmatrix} 1 & 0 & 0 \end{bmatrix} x.$$

First of all, observe that the last two elements of u_1^{con} do absolutely nothing, while the contribution of the last input of u_2^{con} can be absorbed by the first two inputs; hence, in our design, we will turn off those inputs.

The first step is to design a decentralized stabilizing controller for S_{con}. It turns out that this system can be stabilized using static output feedback; indeed, it is easy to verify that the control law

$$u_1^{con} = \begin{bmatrix} 1 \\ 0 \\ 0 \end{bmatrix} y_1^{con} =: J_1 y_1^{con}$$

$$u_2^{con} = \begin{bmatrix} -7.75 \\ -0.25 \\ 0 \end{bmatrix} y_2^{con} =: J_2 y_2^{con}$$

places all three closed-loop eigenvalues at -1.

The second step is to design the generalized-hold-based controller for the plant (5.1) of the form

$$u_i(t) = L_i(t)u_i(kT) = L_i(t)J_i y_i(kT), \ t \in [kT, (k+1)T). \tag{5.11}$$

The quantities J_1 and J_2 have been designed above; we now have to design the generalized-hold $L_i(t)$'s. The standard design requires that

$$\int_0^T e^{A(T-\tau)} B_i L_i(\tau) d\tau = \int_0^T e^{A\tau} B_i L_i(T-\tau) d\tau = T B_i^{con}.$$

The case of $i = 1$ is easy to solve: we can simply set

$$L_1(t) = \begin{bmatrix} 1 \; 0 \; 0 \end{bmatrix}, \ t \in [0, T],$$

so that (5.11) for $i = 1$ becomes

$$u_1(t) = y_1(kT), \ t \in [kT, (k+1)T).$$

The case of $i = 2$ requires more work. First of all, the design can be simplified if we observe that the last element of J_2 is zero, which means that the last element of L_2 is irrelevant: we may as well set it to zero. Hence, we will partition L_2 as

$$L_2(T-\tau) = \begin{bmatrix} \bar{L}_2(\tau) \; 0 \end{bmatrix}$$

and then choose $\bar{L}_2(\tau) \in \mathbf{R}^{1\times 2}$ so that

$$\int_0^T e^{A\tau} B_2 \bar{L}_2(\tau) d\tau = T B_2^{con} \begin{bmatrix} 1\;0 \\ 0\;1 \\ 0\;0 \end{bmatrix}. \tag{5.12}$$

While there are an infinite number of choices of $\bar{L}_2(\tau)$ which satisfy this constraint, for ease of implementation (given that we will typically be implementing this with a zero-order hold), it would be convenient to have this piecewise constant. It turns out that we can choose $\bar{L}_2(\tau)$ to have two values[3] in $\mathbf{R}^{1\times 2}$:

$$\bar{L}_2(\tau) = \begin{cases} g_1 & \tau \in [0, T/2) \\ g_2 & \tau \in [T/2, T). \end{cases} \tag{5.13}$$

If we proceed to solve (5.12) for the unique solution (5.13), we end up with

[3]If we have a general nth order plant, then for most values of T, we can make L_2 have exactly n different values—see Exercise 7.

$$\begin{bmatrix} g_1 \\ g_2 \end{bmatrix} = T \begin{bmatrix} -0.25 + 0.2e^{T/2} + 0.05e^{-2T} & 0.2(e^T - e^{T/2}) + 0.05(e^{-4T} - e^{-2T}) \\ 0.2e^{T/2} - 0.2e^{-2T} & 0.2(e^T - e^{T/2}) - 0.2(e^{-4T} - e^{-2T}) \end{bmatrix}^{-1} \begin{bmatrix} 0 & 1 \\ 1 & -3 \end{bmatrix};$$

with J_2 defined above, we see that the control signal for channel two is given by

$$u_2(t) = \begin{cases} g_2 J_2 y_2(kT) & \tau \in [kT, kT + T/2) \\ g_1 J_2 y_2(kT) & t \in [kT + T/2, (k+1)T). \end{cases}$$

It turns out that the corresponding closed-loop system is stable for $T \in (0, 0.069)$.

For the case of $T = 0.02$, we have carried out a detailed analysis. For this situation, the solution to (5.13) is

$$g_1 = \begin{bmatrix} 3.0204 & -206.1495 \end{bmatrix} \text{ and } g_2 = \begin{bmatrix} 1.0201 & 206.0685 \end{bmatrix}$$

so the controller is

$$u_1(t) = y_1(kT), \; t \in [kT, (k+1)T),$$

$$u_2(t) = \begin{cases} -43.6115 y_2(kT) & \tau \in [kT, kT + T/2) \\ 28.1297 y_2(kT) & t \in [kT + T/2, (k+1)T). \end{cases}$$

With an initial condition of $x(0) = \begin{bmatrix} 1 \\ 1 \\ 1 \end{bmatrix}$, we have carried out a simulation: in Fig. 5.10, we plot x and u. We see that all three states tend exponentially to zero. While u_1 is modest in size, the second control input u_2 is quite large, which is consistent with the size of the time-varying gains used in that channel. We see from this that both x_3 and u_2 are erratic in nature: we provide a close-up of both of these signals for the first 0.1 s in Fig. 5.11. We conclude that while we can achieve the objective of stabilizing the system even though it has an unstable DFM (at $s = 0$), there is a cost: large gains and a rapidly moving control signal, which would require fast actuators. We should not be surprised: after all, there should be a cost to being able to achieve an objective which no LTI controller can provide!

5.6 A Controller Design Based on S_{obs} Using a Generalized Sampler

In this section, we assume that the system has no unstable QFMs and that for every eigenvalue of A there is exactly one Jordan block, so that we can apply Theorem 4 (iii). To this end, let us first form a state-space representation of S_{obs}:

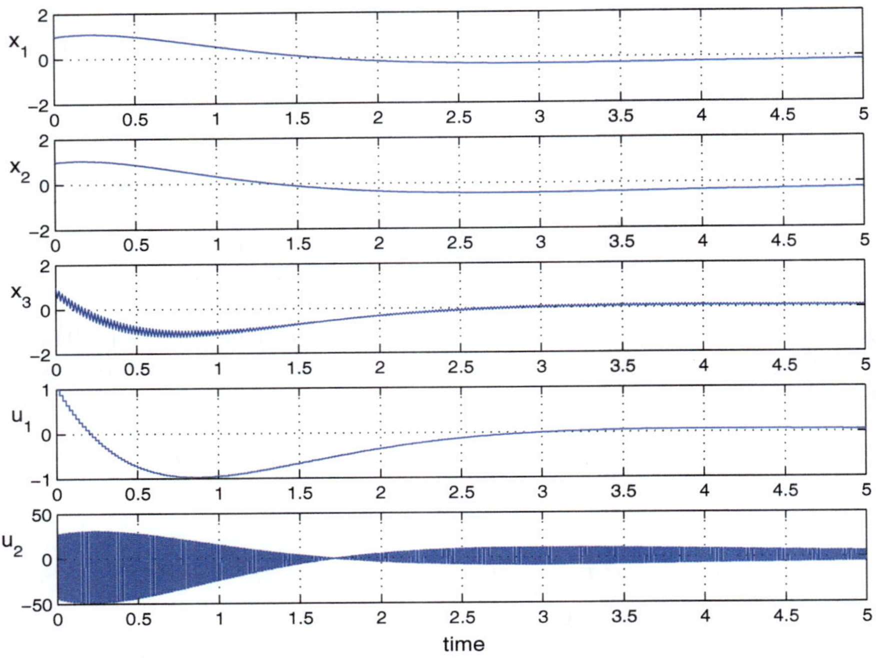

Fig. 5.10 The closed-loop system response with the generalized-hold-based controller applied

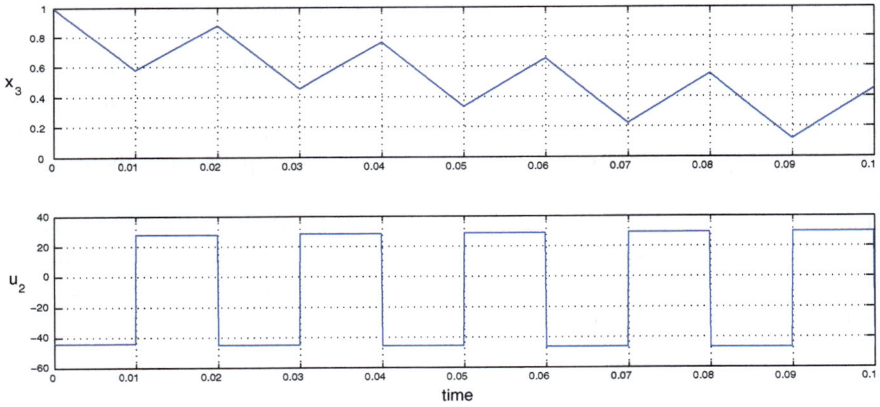

Fig. 5.11 A close-up of the control signal provided by the generalized-hold-based controller

$$\dot{x}^{obs} = Ax^{obs} + \sum_{i=1}^{\nu} B_i u_i^{obs}$$
$$y_i^{obs} = C_i^{obs} x^{obs};$$

in condensed form, this becomes

$$\dot{x}^{obs} = Ax^{obs} + Bu^{obs}$$
$$y^{obs} = C^{obs} x^{obs}.$$

It follows that S_{obs} has no unstable DFMs; this means that there exists an LTI decentralized stabilizing controller K^{obs} for this system:

$$\dot{z}_i^{obs} = F_i z_i^{obs} + G_i y_i^{obs} \tag{5.14}$$
$$u_i^{obs} = H_i z_i^{obs} + J_i y_i^{obs}; \tag{5.15}$$

here, $z_i^{obs}(t) \in \mathbf{R}^{\ell_i}$, and we define $\ell := \ell_1 + \cdots + \ell_p$. Now, let us obtain a concise description of the decentralized stabilizing controller for S_{obs}: with

$$F := \begin{bmatrix} F_1 & & \\ & \ddots & \\ & & F_p \end{bmatrix}, \; G := \begin{bmatrix} G_1 & & \\ & \ddots & \\ & & G_p \end{bmatrix}, \; H := \begin{bmatrix} H_1 & & \\ & \ddots & \\ & & H_p \end{bmatrix},$$

$$J := \begin{bmatrix} J_1 & & \\ & \ddots & \\ & & J_p \end{bmatrix}, \; z^{obs} := \begin{bmatrix} z_1^{obs} \\ \vdots \\ z_p^{obs} \end{bmatrix},$$

it follows that

$$\dot{z}^{obs} = Fz^{obs} + Gy^{obs}$$
$$u^{obs} = Hz^{obs} + Jy^{obs}.$$

If we apply this to S_{obs}, then the fictitious closed-loop system becomes

$$\begin{bmatrix} \dot{x}^{obs} \\ \dot{z}^{obs} \end{bmatrix} = \begin{bmatrix} A + BJC^{obs} & BH \\ GC^{obs} & F \end{bmatrix} \begin{bmatrix} x^{obs} \\ z^{obs} \end{bmatrix} =: A_{cl}^{obs} \begin{bmatrix} x^{obs} \\ z^{obs} \end{bmatrix};$$

clearly, A_{cl}^{obs} is stable.

The goal now is to use the above to design a sampled-data stabilizing controller for the actual system (5.1); the underlying concept used here is the notion of a generalized sampler. The difficulty, of course, is that we cannot measure

$$y_i^{obs}(t) = C_i^{obs} x(t)$$

which is needed to implement the above controller.[4] However, it turns out that we can obtain a good estimate of this quantity, at least periodically. To this end, suppose that

$$u(t) = 0, \ t \in [t_0, t_1].$$

Then

$$y_i(t) = C_i e^{A(t-t_0)} x(t_0), \ t \in [t_0, t_1],$$

so we can, in principle, obtain

$$\begin{bmatrix} y_i^{(1)}(t) \\ y_i^{(2)}(t) \\ \vdots \\ y_i^{(n-1)}(t) \end{bmatrix} = \begin{bmatrix} C_i \\ C_i A \\ \vdots \\ C_i A^{n-1} \end{bmatrix} x(t) = C_i^{obs} x(t), \ t \in [t_0, t_1].$$

Of course, the above requires differentiation, which has undesirable noise properties. We can alleviate this problem to some extent by operating in a sampled-data setting and using an approximation to a differentiator. To this end, we have a **base period** of h, and with $q > n$, we adopt a controller period of $T = qh$. We generate a sequence of estimates $\mathrm{Est}[y_i^{obs}(kT)]$ which we use to drive a discretized version of the above controller K_{obs}, namely

$$z_i((k+1)T) = e^{F_i T} z_i(kT) + [\int_0^T e^{F_i \tau} G_i \, d\tau] \times \mathrm{Est}[y_i^{obs}(kT)] \qquad (5.16)$$

$$u_i(kT) = H_i z_i(kT) + J_i \times \mathrm{Est}[y_i^{obs}(kT)]; \qquad (5.17)$$

the output is used to form $u_i(t)$ in the following way:

$$u_i(t) = \begin{cases} 0 & t \in [kT, kT + (n-1)h) \\ \frac{q}{q-n+1} u_i(kT) & t \in [kT + (n-1)h, (k+1)T). \end{cases} \qquad (5.18)$$

(Notice that the average value of $u_i(t)$ on $[kT, (k+1)T)$ is exactly $u_i(kT)$.) The fact that $u_i(t)$ is set to zero on the first part of every period makes estimating $C_i^{obs} x(kT)$ fairly easy: we have that

$$y_i(t) = C_i e^{A(t-kT)} x(kT), \ t \in [kT, kT + nh],$$

so by using a Taylor series of $e^{A(t-kT)}$ it is straightforward to see that

$$\frac{1}{h}[y_i(kT+h) - y_i(kT)] = C_i A x(kT) + \mathcal{O}(h) x(kT)$$

[4] Here, we have redefined y_i^{obs} slightly to reflect the fact that we are now designing a controller for (5.1) rather than S_{obs}.

as well as

$$\frac{1}{h^2}[y_i(kT+2h) - 2y_i(kT+h) + y_i(kT)] = C_i A^2 x(kT) + \mathcal{O}(h)x(kT).$$

Of course, we can extend this to estimating higher-order derivatives as well. To this end, let us define two $n \times n$ matrices:

$$S := \begin{bmatrix} 1 & 0 & 0 & \cdots & 0 \\ 1 & 1 & 1 & \cdots & 1 \\ 1 & 2 & 2^2 & \cdots & 2^{n-1} \\ & & \vdots & & \\ 1 & n-1 & (n-1)^2 & \cdots & (n-1)^{n-1} \end{bmatrix}, \quad H(h) := \mathrm{diag}\{1, h, \frac{h^2}{2!}, \ldots, \frac{h^{n-1}}{(n-1)!}\}.$$

With $I_j \in \mathbf{R}^{r_j \times r_j}$ the identity matrix of dimension r_j, we also define the square invertible matrices

$$S_j := S \otimes I_j, \quad H_j := H \otimes I_j, \quad j = 1, \ldots, \nu.$$

We also define a list of sampled output signals:

$$\mathcal{Y}_i(t) := \begin{bmatrix} y_i(t) \\ y_i(t+h) \\ \vdots \\ y_i(t+(n-1)h) \end{bmatrix}.$$

Lemma 1 *There exists a constant $\gamma > 0$ so that if the control signal is given by (5.18), then the solution of (5.1) satisfies, for $i = 1, \ldots, \nu$:*

$$\|H_i(h)^{-1} S_i^{-1} \mathcal{Y}_i(kT) - C_i^{obs} x(kT)\| \le \gamma h \|x(kT)\|;$$

furthermore,

$$\|x((k+1)T) - e^{AT}x(kT) - T\sum_{j=1}^{\nu} B_j u_j(kT)\| \le \gamma h^2 (\|x(kT)\| + \|u(kT)\|).$$

Proof The first part is proven in Lemma 1 of [12] for the case of all channels being scalar; the case of non-scalar channels follows easily. The second bound is straightforward to derive by solving the state equation and using Taylor series.

□

This lemma provides a clear choice for the estimate of $C_i^{obs}x(kT)$ which we label $\text{Est}[y_i^{obs}(kT)]$:

$$\text{Est}[y_i^{obs}(kT)] := H_i(h)^{-1}S_i^{-1}\mathcal{Y}_i(kT),\ i = 1, \ldots, \nu. \tag{5.19}$$

So the controller consists of the discrete-time compensator (5.16)–(5.17), the control signal (5.18) and the estimate of $C_i^{obs}x(kT)$, given above by (5.19), which is used to drive the discrete-time compensator (5.16)–(5.17). Given that the controller state is in discrete time while the plant state is in continuous time, we define the sampled-data state by

$$x_{sd}(t) := \begin{bmatrix} x(t) \\ z[k] \end{bmatrix},\ t \in [kT, (k+1)T).$$

Theorem 6 *With $q > n$, suppose that the controller given by (5.16)–(5.17), (5.18) and (5.19) is applied to the plant (5.1) at time $t_0 = 0$. Then there exist constants $c > 0$, $\bar{T} > 0$ and $\lambda < 0$ so that for every $T \in (0, \bar{T})$, $x_0 \in \mathbf{R}^n$ and $z(0) \in \mathbf{R}^\ell$, we have*

$$\|x_{sd}(t)\| \leq ce^{\lambda t}\|x_{sd}(0)\|,\ t \geq 0.$$

Remark 4 Since the controller is linear periodic, it is easy to show that if we add noise at the plant controller interface, then its effect on x_{sd} has a bounded gain in the induced ∞-norm sense.

Proof Fix $q > n$, and let $x_0 \in \mathbf{R}^n$, $z(0) \in \mathbf{R}^\ell$ and $T > 0$ be arbitrary. If we apply the proposed controller to the model of S_{obs} and apply Lemma 1, then we obtain

$$\begin{aligned}
\begin{bmatrix} x((k+1)T) \\ z((k+1)T) \end{bmatrix} &= \begin{bmatrix} e^{AT}x(kT) + TBu(kT) + \mathcal{O}(T^2)(\|x(kT)\| + \|u(kT)\|) \\ e^{FT}z(kT) + [\int_0^T e^{F\tau}G\,d\tau] \times [C^{obs}x(kT) + \mathcal{O}(T)x(kT)] \end{bmatrix} \\
&= \begin{bmatrix} (I + AT + BJC^{obs}T)x(kT) + TBHz(kT) + \mathcal{O}(T^2)(\|x(kT)\| + \|z(kT)\|) \\ (I + FT)z(kT) + TGC^{obs}x(kT) + \mathcal{O}(T^2)(\|x(kT)\| + \|z(kT)\|) \end{bmatrix} \\
&= [I + A_{cl}^{obs}T]\begin{bmatrix} x(kT) \\ z(kT) \end{bmatrix} + \mathcal{O}(T^2)(\|x(kT)\| + \|z(kT)\|).
\end{aligned}$$

Since A_{cl}^{obs} is stable, there exists a positive definite and symmetric square matrix P which satisfies

$$[A_{cl}^{obs}]^T P + PA_{cl}^{obs} + I = 0;$$

define the Lyapunov function

$$V(x_{sd}(kT)) := x_{sd}(kT)^T Px_{sd}(kT).$$

It is easy to see that

$$\begin{aligned} V(x_{sd}((k+1)T)) &= V(x_{sd}(kT)) - T\|x_{sd}(kT)\|^2 + \mathcal{O}(T^2)\|x_{sd}(kT)\|^2 \\ &\leq V(x_{sd}(kT)) - T\frac{1}{\|P\|}V(x_{sd}(kT) + \mathcal{O}(T^2)V(x_{sd}(kT); \end{aligned}$$

for small T, say $T \in (0, \bar{T})$, we have

$$V(x_{sd}((k+1)T)) \leq [1 - T\frac{1}{2\|P\|}]V(x_{sd}(kT)),$$

which means that

$$V(x_{sd}(kT)) \leq e^{-kT\frac{1}{2\|P\|}} V(x_{sd}(0)), \; k \geq 0.$$

If we set

$$c_1 := [\frac{\|P\|}{\lambda_{min}(P)}]^{1/2}$$

and $\lambda := -\frac{1}{4\|P\|}$, we see that for $T \in (0, \bar{T})$:

$$\|x_{sd}(kT)\| \leq c_1 e^{\lambda kT}\|x(0)\|, \; k \geq 0.$$

It follows from Lemma 1 that there is a bound on the inter-sample behaviour: for small enough $T \in (0, \bar{T})$, we see that

$$\|x_{sd}(t)\| \leq 2c_1 e^{\lambda t}\|x_{sd}(0)\|, \; t \geq 0.$$

□

Example 8 Now we illustrate the design procedure proposed here. Consider the plant

$$\begin{aligned} \dot{x} &= \begin{bmatrix} 0 & 0 & 0 \\ 0 & 0 & 1 \\ 0 & 4 & -3 \end{bmatrix} x + \begin{bmatrix} 1 \\ 0 \\ 0 \end{bmatrix} u_1 + \begin{bmatrix} 0 \\ 0 \\ 1 \end{bmatrix} u_2 \\ y_1 &= \begin{bmatrix} 0 & 0 & 1 \end{bmatrix} x \\ y_2 &= \begin{bmatrix} 1 & 0 & 0 \end{bmatrix} x, \end{aligned}$$

with a corresponding transfer function of

$$\begin{bmatrix} Y_1(s) \\ Y_2(s) \end{bmatrix} = \begin{bmatrix} 0 & \frac{s}{(s-1)(s+4)} \\ \frac{1}{s} & 0 \end{bmatrix} \begin{bmatrix} U_1(s) \\ U_2(s) \end{bmatrix}.$$

From the transfer function, we see that there is a DFM at $s = 0$; however, since the system is centrally controllable and centrally observable, and the associated graph is clearly strongly connected, it follows from Theorem 2 that it has no QFMs; from Theorem 1, we see that there is linear time-varying stabilizing controller. It can be easily checked that the discretized version of this plant has an unstable DFM at $z = 1$, so we cannot simply use a simple sampled-data controller considered in Sect. 5.2. However, it is easy to check that A has distinct eigenvalues, so Theorem 4(iii) says that the associated system S_{obs} has no DFMs, which means that the procedure adopted in this section can be employed to design a linear periodic stabilizing controller.

To this end, we first form the associated model S_{obs}: we have

$$\dot{x}^{obs} = \begin{bmatrix} 0 & 0 & 0 \\ 0 & 0 & 1 \\ 0 & 4 & -3 \end{bmatrix} x^{obs} + \begin{bmatrix} 1 \\ 0 \\ 0 \end{bmatrix} u_1^{obs} + \begin{bmatrix} 0 \\ 0 \\ 1 \end{bmatrix} u_2^{obs}$$

$$y_1^{obs} = \begin{bmatrix} 0 & 0 & 1 \\ 0 & 4 & -3 \\ 0 & -12 & 13 \end{bmatrix} x^{obs}$$

$$y_2^{obs} = \begin{bmatrix} 1 & 0 & 0 \\ 0 & 0 & 0 \\ 0 & 0 & 0 \end{bmatrix} x^{obs},$$

First of all, observe that the contribution of the last output of y_1^{obs} is a linear combination of the first two, so if we design our controller carefully, then we will not need to estimate that term. Second of all, the last two elements of y_2^{obs} are zero, so we clearly need not estimate them either.

The first step is to design a decentralized stabilizing controller for S_{obs}. It turns out that this system can be stabilized using static output feedback; indeed, it is easy to verify that the control law

$$u_1^{obs} = \begin{bmatrix} -7.75 & -0.25 & 0 \end{bmatrix} y_1^{obs} =: J_1 y_1^{obs}$$

$$u_2^{obs} = \begin{bmatrix} 1 & 0 & 0 \end{bmatrix} y_2^{obs} =: J_2 y_2^{obs}$$

places all three closed-loop eigenvalues at -1. Notice that the gain on the last element of y_1^{obs} and the last two elements of y_2^{obs} is zero.

The second step is to design the generalized sampler at each output. In channel one, we can measure the first elements on $y_1^{obs}(t)$—it is simply $y_1(t)$. The second element can be estimated at time kT via $\frac{1}{h}[y_1(kT + h) - y_1(kT)]$; this is a more efficient procedure to estimate this quantity that the one provided in (5.19). We do not need to estimate the third element since it is not used, so we end up with

$$\text{Est}[y_1^{obs}(kT)] := \begin{bmatrix} y_1(kT) \\ \frac{1}{h}[y_1(kT + h) - y_1(kT)] \\ 0 \end{bmatrix}.$$

In channel two, we only need to estimate the first element to y_2^{obs} since the last two elements are not used; we end up with

$$\text{Est}[y_2^{obs}(kT)] := \begin{bmatrix} y_2(kT) \\ 0 \\ 0 \end{bmatrix}.$$

Here, we only required two samples of y to obtain the estimates, which means that the estimation part only requires $(n-2)h = h$ units of time rather than $(n-1)h = 2h$ units of time, so Eq. (5.18) is adjusted accordingly by replacing $\frac{q}{q-n+1}$ by $\frac{q}{q-n+2} = \frac{q}{q-1}$. For the case of $h = 0.01$, $q = 2$ and $T = 0.02$, we have carried out a detailed analysis. For this situation, the controller is

$$\begin{aligned} u_1(t) &= \begin{cases} 0 & t \in [kT, kT+h) \\ 2J_1 \text{Est}[y_1^{obs}(kT)] & t \in [kT+h, (k+1)T) \end{cases} \\ &= \begin{cases} 0 & t \in [kT, kT+h) \\ \begin{bmatrix} -15.5 & -0.5 \end{bmatrix} \begin{bmatrix} y_1(kT) \\ \frac{1}{h}[y_1(kT+h) - y_1(kT)] \end{bmatrix} & t \in [kT+h, (k+1)T) \end{cases} \\ &= \begin{cases} 0 & t \in [kT, kT+h) \\ 34.5 y_1(kT) - 50 y_1(kT+h) & t \in [kT+h, (k+1)T) \end{cases} \end{aligned}$$

together with

$$\begin{aligned} u_2(t) &= \begin{cases} 0 & t \in [kT, kT+h) \\ 2J_2 \text{Est}[y_2^{obs}(kT)] & t \in [kT+h, (k+1)T) \end{cases} \\ &= \begin{cases} 0 & t \in [kT, kT+h) \\ 2y_2(kT) & t \in [kT+h, (k+1)T). \end{cases} \end{aligned}$$

With an initial condition of $x(0) = \begin{bmatrix} 1 \\ 1 \\ 1 \end{bmatrix}$, we have carried out a simulation: in Fig. 5.12, we plot x and u. We see that all three states tend exponentially to zero. While u_1 and u_2 are modest in size, they move rapidly; see Fig. 5.13 for a close-up during the first 0.1 s. We conclude that while we can achieve the objective of stabilizing the system even though it has an unstable DFM (at $s = 0$), there is a cost: a rapidly moving control signal, which would require fast actuators.[5] We should not be surprised: after all, there should be a cost to being able to achieve an objective which no LTI controller can provide!

[5] Here, we avoid the use of large control signals, a problem that is associated with the generalized-hold approach of the previous section.

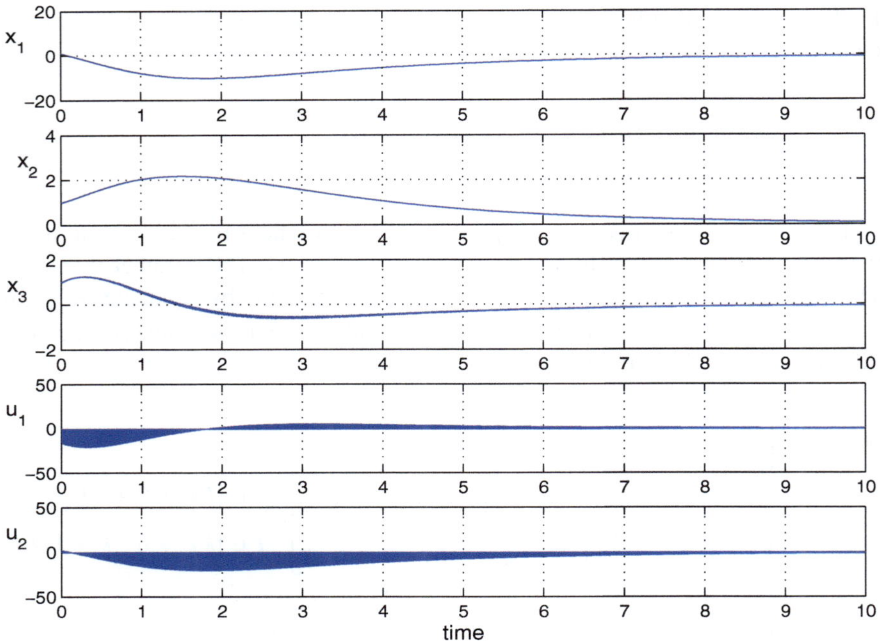

Fig. 5.12 The closed-loop system response with the generalized sampler-based controller applied

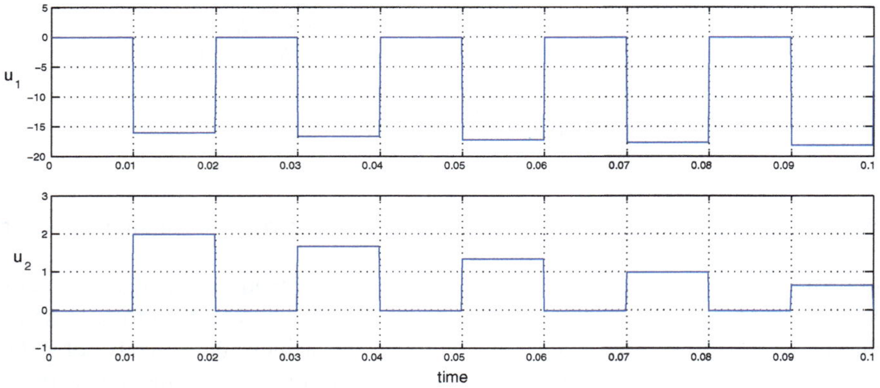

Fig. 5.13 A close-up of the controller signal when the generalized sampler-based controller is applied

5.7 A Controller Design Based on S_{mar}

In this section,[6] we assume that the system has no unstable QFMs and that for every eigenvalue of A there may be more than one Jordan block, so that we can apply Theorem 4 (i). To this end, let us first form a state-space representation of S_{mar}:

$$\begin{aligned} \dot{x}^{mar} &= Ax^{mar} + \sum_{i=1}^{\nu} B_i^{mar} u_i^{mar} \\ y_i^{mar} &= C_i^{mar} x^{mar} + \sum_{j=1}^{\nu} D_{ij}^{mar} u_j^{mar}; \end{aligned}$$

in condensed form, this becomes

$$\begin{aligned} \dot{x}^{mar} &= Ax^{mar} + B^{mar} u^{mar} \\ y^{mar} &= C^{mar} x^{mar} + D^{mar} u^{mar}. \end{aligned}$$

It follows that S_{mar} has no unstable DFMs; this means that there exists an LTI decentralized stabilizing controller K^{mar} for this system, and it is easy to confirm that it does not need a direct feedthrough term:

$$\left. \begin{aligned} \dot{z}_i^{mar} &= F_i z_i^{mar} + G_i y_i^{mar}, \\ u_i^{mar} &= H_i z_i^{mar}, \quad i = 1, \ldots, \nu; \end{aligned} \right\} \tag{5.20}$$

here, $z_i^{mar}(t) \in \mathbf{R}^{\ell_i}$, and we define $\ell := \ell_1 + \cdots + \ell_\nu$. It will be convenient to group the inputs and outputs, which motivates the following definitions:

$$z^{mar} = \begin{bmatrix} z_1^{mar} \\ \vdots \\ z_p^{mar} \end{bmatrix}, \; F = \begin{bmatrix} F_1 & & \\ & \ddots & \\ & & F_p \end{bmatrix}, \; G = \begin{bmatrix} G_1 & & \\ & \ddots & \\ & & G_p \end{bmatrix}, \; H = \begin{bmatrix} H_1 & & \\ & \ddots & \\ & & H_p \end{bmatrix},$$

$$y^{mar} = \begin{bmatrix} y_1^{mar} \\ \vdots \\ y_p^{mar} \end{bmatrix}, \quad u^{mar} = \begin{bmatrix} u_i^{mar} \\ \vdots \\ u_p^{mar} \end{bmatrix}.$$

In closed loop, we have a state-space description given by

$$\begin{bmatrix} \dot{x}^{mar} \\ \dot{z}^{mar} \end{bmatrix} := \underbrace{\begin{bmatrix} A & B^{mar} H \\ G C^{mar} & F + G D^{mar} H \end{bmatrix}}_{=:A_{cl}^{mar}} \underbrace{\begin{bmatrix} x^{mar} \\ z^{mar} \end{bmatrix}}_{=:x_{cl}^{mar}}; \tag{5.21}$$

[6]This section is derived, in part, from [16].

clearly, A_{cl}^{mar} is stable.

The goal here is to design a controller for the actual plant (5.1) rather than for S_{mar}. The design will be much more involved than that used in the previous two sections; we will provide only a high-level description. We use a controller which consists of two parts: a suitably modified version of the LTI compensator K^{mar} together with a linear periodic sampled-data component **of period** T which carries out periodic probing and estimation: in closed loop, the system behaviour is similar to that governed by (5.21). To make this work, there are two key ideas:

(i) First, let us redefine y_i^{mar} slightly to reflect the fact that the plant state is x and the control input is u:

$$y_i^{mar} = C_i^{mar} x + \sum_{j=1}^{\nu} D_{ij}^{mar} u_j;$$

since this quantity is not directly measurable in channel i, we will periodically generate an estimate $\hat{y}_i^{mar}(kT)$. We carry out the estimation over the first part of the period $[kT, kT + T')$; we achieve this by probing the plant in such a way that the overall effect of the probing has a second-order effect on the plant state.

(ii) We then use the estimate $\hat{y}_i^{mar}(kT)$ to drive a suitably modified version of K_{mar}, which, in turn, generates the control signal: in channel i, we have

$$\dot{z}_i(t) = F_i z_i(t) + \begin{cases} 0 & t \in [kT, kT + T') \\ \frac{T}{T-T'} G_i^{mar} \hat{y}_i^{mar}(kT) & t \in [kT + T', (k+1)T), \end{cases}$$
$$\hat{u}_i^{mar}(t) = H_i z_i(t), t \in [kT, (k+1)T).$$

During the first part of the period, u_i is used to carry out the probing; given that B_i^{mar} is simply B_i padded with zeros, it is reasonable that on the remainder of the period we set

$$u_i(t) = \frac{T}{T - T'} \begin{bmatrix} I_{m_i} & 0 \end{bmatrix} \hat{u}_i^{mar}(t), \ t \in [kT + T', (k+1)T).$$

If the estimate $\hat{y}_i^{mar}(kT)$ is accurate, and the probing is done in such a way that it leaves only a second-order affect on the state, then it is reasonable to expect that for small $T > 0$ we have

$$\begin{bmatrix} x((k+1)T) \\ z((k+1)T) \end{bmatrix} \approx [I + A_{cl}^{mar} T] \begin{bmatrix} x(kT) \\ z(kT) \end{bmatrix},$$

which means that we achieve closed-loop stability if T is small enough.

The remaining issue to discuss is that of estimating $\hat{y}_i^{mar}(kT)$ on $[kT, kT + T')$ in channel i. First of all, since C_i^{mar} is just C_i padded with zeros, it is clear that we can form $C_i^{mar}x(kT)$ simply from the measurement $y_i(kT)$. Generating a good estimate of $\sum_{j=1}^{\nu} D_{ij}^{mar}u_j(t)$ in channel i is much more difficult. In channel i, we generate $u_i(t)$ so we can clearly construct $D_{ii}^{mar}u_i(kT)$; we will carry out probing and estimation on the plant in order to estimate the remaining terms in the sum. There are many ways to carry out probing and estimation to generate this estimate; below, we will provide one of them. Suppose that we turn off all of the plant inputs on the interval $[kT, kT + T_1)$; then

$$y_i(t) = C_i e^{A(t-kT)}x(kT), \ t \in [kT, kT + T_1);$$

clearly, $y_i(t)$ on the interval $[kT, kT + T_1)$ has sufficient information to determine $C_i^{obs}x(kT)$ in channel i.[7] **Now suppose that all channels are scalar valued**, and we apply a probing signal $\delta \neq 0$ of the form

$$u(t) = e_j\delta, \ t \in [kT + T_1, kT + T_2);$$

we see that

$$y_i(t) = C_i e^{A(t-kT)}x(kT) + \int_{kT+T_1}^{t} C_i e^{A(t-\tau)}B_j\delta, \ t \in [kT + T_1, kT + T_2).$$

Knowledge of $C_i^{obs}x(kT)$ obtained previously is enough to ascertain the first term on the RHS, which means that

$$\int_{kT+T_1}^{t} C_i e^{A(t-\tau)}B_j\delta, \ t \in [kT + T_1, kT + T_2),$$

can be measured in channel i; as long as the transfer function $C_i(sI - A)^{-1}B_j$ is not identically zero, it is clear that δ can be estimated in channel i, i.e. it is possible to pass information from channel j to channel i. (If $C_i(sI - A)^{-1}B_j$ is identically zero, then $D_{ij}^{mar} = 0$, so we do not actually need to estimate δ.) Of course, this can be repeated in every channel, so that with a finite amount of probing on an interval of non-zero length, say $[kT, kT + T')$, we can obtain an accurate estimate of $\hat{y}_i^{mar}(kT)$ in channel i. Hence, it is intuitively reasonable that we can achieve our objective of designing a stabilizing controller in the above fashion.

At this point, we refer you to the paper [16] for a detailed controller design, which yields a periodic sampled-data stabilizing controller; the estimation approach there is slightly different than the above. A proof is provided, as well as a detailed design example.

[7] Recall that C_i^{obs} is simply the observability matrix of the original plant associated with channel i.

5.8 Summary

The idea that nonlinear and/or time-varying controllers do better than LTI ones in the decentralized context goes back at least to Kobyashi et al. [11], wherein a study of controllability in the open-loop decentralized setting is carried out. This was one of the motivations for the work of Anderson and Moore [2], wherein it was proven that some unstable DFMs can be tolerated when using linear time-varying feedback; at about the same time, Wang [18] argued that a generalized-hold-based controller can achieve a similar result in some situations. This lead to further work on this topic, both in continuous time, e.g. see Willems [19], and discrete time, e.g. see Khargonekar and Ozguler [10], as well as an attempt to classify DFMs into those which are, in some sense, truly fixed, and those which are "moveable" using the appropriate NLTV controller. There are three key results along this line:

(i) In Anderson and Moore [2], the case of a centrally controllable and observable system is considered; as discussed in Sect. 5.3, they argue that if the graph associated with the system is strongly connected, then there always exists a time-varying controller to provide closed-loop stability, even if the system has unstable DFMs.
(ii) In Ozguner and Davison [17], the class of decentralized systems with distinct and nonzero eigenvalues is considered, and the notion of *structured DFMs* (SDFMs) is defined; once the A matrix is written in diagonal form, the SDFMs are those DFMs which remain as DFMs after perturbing the nonzero elements in the system input/output matrices. Their result is constructive: if the system has no unstable SDFMs, and a sampler is added to the plant output and a zero-order hold to the plant input, then the resulting discretized plant has no unstable DFMs, so a discrete-time compensator can be used to stabilize the plant.[8] As discussed in Sect. 5.2, the approach may fail if the system has an eigenvalue at zero.
(iii) As discussed in Sect. 5.3, Gong and Aldeen [7] use graph theory and the notion of a quotient system to introduce the notion of a QFM, and demonstrate that there exists a decentralized NLTV stabilizing controller if and only if the system has no unstable QFMs. It turns out that for the class of systems considered in [17], the notion of QFM and SDFM are one and the same.

More recently, two of the co-authors of this book became interested in finding a non-graph-centric method to compute the QFMs [14]; the results are discussed in Sect. 5.4. After that, their attention turned to controller synthesis [16], which is discussed in Sect. 5.7. Because of the complexity of that result, some simpler synthesis methods are provided in Sects. 5.5 and 5.6; the idea of using generalized holds in decentralized control has been around for some time, e.g. see Wang [18] as well as Aghdam and Davison [1].

---

[8] The overall controller is sampled-data: it consists of a sampler, a zero-order hold and the LTI discrete-time compensator, which together comprise a linear periodic controller.

5.9 Problems

1. In Sect. 5.2, we considered an example and showed that the discretized system has a zero at

$$f(T) = \frac{4e^{2T} - 3e^{T} - e^{-T}}{e^{2T} - 4e^{-T} + 3}.$$

Verify that

$$f(T) = e^{T} - \frac{1}{6}T^3 + \mathcal{O}(T^4).$$

2. For the system given in Remark 1, find the discretized version of the system (when using a zero-order hold), and then verify that it has a DFM at $z = 1$ for every sampling period $T > 0$.
3. Consider the decentralized system

$$\dot{x} = \begin{bmatrix} 1 & 0 & 0 \\ 0 & 2 & 0 \\ 0 & 0 & 3 \end{bmatrix} x + \begin{bmatrix} 0 \\ 1 \\ 1 \end{bmatrix} u_1 + \begin{bmatrix} 1 \\ 0 \\ 0 \end{bmatrix} u_2$$
$$y_1 = \begin{bmatrix} 1 & 0 & 0 \end{bmatrix} x$$
$$y_2 = \begin{bmatrix} 0 & -1 & 2 \end{bmatrix} x.$$

Verify that this system has an unstable DFM but the discretized version does not (for any $T > 0$).

4. Consider the following system:

$$\dot{x} = \begin{bmatrix} 1 & 0 & 0 \\ 0 & 1 & 0 \\ 0 & 0 & 1 \end{bmatrix} x + \begin{bmatrix} 0 & 0 \\ 1 & 0 \\ 0 & 1 \end{bmatrix} u_1 + \begin{bmatrix} 1 \\ 0 \\ 0 \end{bmatrix} u_2$$
$$y_1 = \begin{bmatrix} 1 & 0 & 0 \end{bmatrix} x$$
$$y_2 = \begin{bmatrix} 0 & 1 & 0 \\ 0 & 0 & 1 \end{bmatrix} x.$$

(a) Construct the graph of the system and verify that it is strongly connected.
(b) Find the systems DFMs and verify that it has an unstable one.
(c) Find the discretized version of this system and verify that it has an unstable DFM for every sampling period $T > 0$.
(d) Construct the quotient system and find its DFMs (the QFMs of the original system).
(e) Form the model S_{con} and find its DFMs. Verify that they are different from the QFMs of the original system.
(f) Form the model S_{obs} and find its DFMs. Verify that they are different from the QFMs of the original system.

(g) Form the model S_{mar} and find its DFMs. Verify that they are equal to the QFMs of the original system.

5. In the graphical set-up of Sect. 5.2, we showed how inputs and outputs can be regrouped to form a quotient system. As shown there, the corresponding topology is upper block triangular, with dimensions consistent with the quotient system partitioning. Now, show that the state-space model can be transformed in such a way that it also admits an upper block triangular representation in the A, B and C matrices. (Hint: Use the approach of Corfmat and Morse [3].)
6. In designing generalized holds, an important step is to find a time-varying function of time $L_i(t)$ so that

$$\int_0^T e^{A(T-\tau)} B_i L_i(\tau)\, d\tau$$

equals any desired vector in the image of the controllability matrix

$$\left[\, B_i \; AB_i \; \cdots \; A^{n-1}B_i \,\right].$$

Provide a constructive proof of this statement.
7. Show that, in the previous question, for most $T > 0$, we can choose $\bar{n} \in \{1, \ldots, n\}$ and design $L_i(t)$ so that it is piecewise constant and of the form

$$L_i(t) = \begin{cases} l_1 & t \in [0, \frac{T}{\bar{n}}) \\ l_2 & t \in [\frac{T}{\bar{n}}, 2\frac{T}{\bar{n}}) \\ \vdots & \vdots \\ l_{\bar{n}} & t \in [(\bar{n}-1)\frac{T}{\bar{n}}, T). \end{cases} \tag{5.22}$$

Hint: First of all, set $\bar{n}$ equal to the rank of the controllability matrix

$$\left[\, B_i \; AB_i \; \cdots \; A^{n-1}B_i \,\right].$$

Second of all, you may use, without proof, the fact from [6] that if the pair $(\bar{A}, \bar{B}) \in \mathbf{R}^{\bar{n}\times\bar{n}} \times \mathbf{R}^{\bar{n}\times m_i}$ is controllable, then for most $T > 0$, the pair $(e^{\bar{A}T/\bar{n}}, \int_0^{T/\bar{n}} e^{\bar{A}\tau}\bar{B}\, d\tau)$ is controllable.

References

1. A.G. Aghdam, E.J. Davison, R. Becerril, Structural modification of systems using discretization and generalized sampled-data holds. Automatica **42**, 1935–1941 (2006)
2. B.D.O. Anderson, J.B. Moore, Time-varying feedback laws for decentralized control. IEEE Trans. Autom. Control **AC–26**, 1133–1139 (1981)

3. J.P. Corfmat, A.S. Morse, Decentralized control of linear multivariable systems. Automatica **12**, 479–495 (1976)
4. E.J. Davison, T.N. Chang, Decentralized stabilization and pole assignment for general proper systems. IEEE Trans. Autom. Control **AC–35**, 652–664 (1990)
5. E.J. Davison, S.H. Wang, On the stabilization of decentralized control systems. IEEE Trans. Autom. Control **5**, 473–478 (1973)
6. B.A. Francis, T.T. Georgiou, Stability theory for linear time-invariant plants with periodic digital controllers. IEEE Trans. Autom. Control **AC–33**, 820–832 (1988)
7. Z. Gong, M. Aldeen, Stabilization of decentralized control systems. Math. Syst. Estimation Control **7**, 1–16 (1997)
8. M. Ikeda, H. Maeda, S. Kodama, Estimation and feedback in linear time-varying systems: a deterministic theory. SIAM J. Control **13**(2), 304–326 (1975)
9. P.T. Kabamba, Control of linear systems using generalized sampled-data hold functions. IEEE Trans. Autom. Control. **AC–32**, 772–783 (1987)
10. P.K. Khargonekar, A.B. Ozguler, Decentralized control and periodic feedback. IEEE Trans. Autom. Control **AC–39**, 877–882 (1994)
11. H. Kobyashi, H. Hanafusa, T. Yoshikawa, Controllability under decentralized information structure. IEEE Trans. Autom. Control **AC–23**(2), 182–188 (1978)
12. D.E. Miller, Near optimal LQR performance for a compact set of plants. IEEE Trans. Autom. Control **51**, 1423–1439 (2006)
13. D.E. Miller, E.J. Davison, On computing quotient decentralized fixed modes, in *Proceedings of the 18th IFAC World Congress* (Milan, Italy, 2011), pp. 2546–2551
14. D.E. Miller, E.J. Davison, An algebraic characterization of quotient decentralized fixed modes. Automatica **48**, 1639–1644 (2012)
15. D.E. Miller, E.J. Davison, Near optimal control in the decentralized setting. IEEE Trans. Autom. Control **59**(2), 327–340 (2014)
16. D.E. Miller, E.J. Davison, Linear periodic controller design for decentralized control systems. Syst. Control Lett. **64**, 1–11 (2014)
17. U. Ozguner, E.J. Davison, Sampling and decentralized fixed modes, in *Proceedings of the American Control Conference* (1985), pp. 257–262
18. S.H. Wang, Stabilization of decentralized control systems via time-varying controllers. IEEE Trans. Autom. Control **AC–27**, 741–749 (1982)
19. J.L. Willems, Time-varying feedback for the stabilization of fixed modes in decentralized control. Automatica **25**, 127–131 (1989)

Chapter 6
Optimal Decentralized Control

6.1 Quadratic Performance Index

Consider the following system:

$$\begin{aligned}\dot{x}(t) &= Ax(t) + \sum_{i=1}^{\nu} B_i u_i(t) + E\omega(t) \\ y_i(t) &= C_i x(t) + D_i u_i(t) + F_i \omega(t), \quad i = 1, \ldots, \nu \\ y_i^{\mathrm{m}}(t) &= C_i^{\mathrm{m}} x(t) + D_i^{\mathrm{m}} u_i(t) + F_i^{\mathrm{m}} \omega(t), \quad i = 1, \ldots, \nu \end{aligned} \tag{6.1}$$

where the vector variables and matrix parameters in the above equations are the same as those given in (3.1). It is desired to design a decentralized controller with the diagonal information flow structure **K** given below:

$$\mathbf{K} := \left\{ K \in R^{m \times r} | K = \begin{bmatrix} K_1 & & \mathbf{0} \\ & \ddots & \\ \mathbf{0} & & K_\nu \end{bmatrix}, \; K_i \in R^{m_i \times r_i}, \; i = 1, \ldots, \nu \right\}$$

to stabilize the system and achieve a good transient and steady-state response. We will need to (i) check the existence conditions of a controller to solve the robust decentralized servomechanism problem; (ii) define a performance index which reflects the transient behaviour of the system output and (iii) design a decentralized controller to minimize the above performance index.

In centralized optimal control problem, usually a quadratic performance index of the following form:

$$J = \int_0^\infty \left(x'(t)\breve{Q}x(t) + u'(t)\breve{R}u(t) \right) \mathrm{d}t$$

E. J. Davison et al., *Decentralized Control of Large-Scale Systems*,
https://doi.org/10.1007/978-1-4419-6014-6_6

is considered, which represents a combination of the energy of the system state and the energy of the control input to be balanced against each other. One of the advantages of this performance index is that the optimal solution is in the form of state feedback, whose gain can be obtained analytically by solving the algebraic Riccati equation written in terms of the system matrices. The minimum performance index will depend on the initial conditions. However, the optimal control parameters will be independent of the initial conditions.

We will use the quadratic performance index for the decentralized optimal control problem as well. However, since the states of the system are not available (and cannot be estimated in each local output station by using an observer, in general), it is important to use an output-based performance index in terms of the measured error of the local outputs. To this end, let the following quadratic function be used:

$$J = \int_0^{\infty} \left(e'(t)\breve{Q}e(t) + u'(t)\breve{R}u(t) \right) \mathrm{d}t \tag{6.2}$$

6.2 Decentralized Control Structure

Consider the following decentralized controller:

$$\begin{aligned} \dot{z}_i(t) &= S_i z_i(t) + R_i e_i(t) \\ u_i(t) &= Q_i z_i(t) + K_i y_i^{\mathrm{m}}(t), \quad i = 1, \ldots, \nu \end{aligned} \tag{6.3}$$

where $e_i(t) = y_i(t) - y_i^{\mathrm{ref}}(t),\ i = 1, \ldots, \nu$, and the initial controller states are assumed to be zero, i.e. $z_i(0) = 0,\ i = 1, \ldots, \nu$. Assume also that this controller solves the robust decentralized servomechanism problem for a given class of reference and disturbance inputs. For simplicity and without loss of generality, let $D_i = 0,\ i = 1, \ldots, \nu$. The augmented system equations can be written as follows:

$$\begin{aligned} \begin{bmatrix} \dot{x}(t) \\ \dot{z}(t) \end{bmatrix} &= \begin{bmatrix} A + BKC_{\mathrm{m}} & BQ \\ RC & S \end{bmatrix} \begin{bmatrix} x(t) \\ z(t) \end{bmatrix} + \begin{bmatrix} 0 \\ -R \end{bmatrix} y_{\mathrm{ref}}(t) + \begin{bmatrix} E \\ RF \end{bmatrix} \omega(t) \\ y(t) &= \begin{bmatrix} C & 0 \end{bmatrix} \begin{bmatrix} x(t) \\ z(t) \end{bmatrix}, \\ u(t) &= \begin{bmatrix} KC_{\mathrm{m}} & Q \end{bmatrix} \begin{bmatrix} x(t) \\ z(t) \end{bmatrix}, \end{aligned} \tag{6.4}$$

where

$$S := \begin{bmatrix} S_1 & & \mathbf{0} \\ & \ddots & \\ \mathbf{0} & & S_\nu \end{bmatrix}, R := \begin{bmatrix} R_1 & & \mathbf{0} \\ & \ddots & \\ \mathbf{0} & & R_\nu \end{bmatrix},$$

$$Q := \begin{bmatrix} Q_1 & & \mathbf{0} \\ & \ddots & \\ \mathbf{0} & & Q_\nu \end{bmatrix}, K := \begin{bmatrix} K_1 & & \mathbf{0} \\ & \ddots & \\ \mathbf{0} & & K_\nu \end{bmatrix},$$

$$B := \begin{bmatrix} B_1 \cdots B_\nu \end{bmatrix}, C := \begin{bmatrix} C_1 \\ \vdots \\ C_\nu \end{bmatrix}, C_{\text{m}} := \begin{bmatrix} C_1^{\text{m}} \\ \vdots \\ C_\nu^{\text{m}} \end{bmatrix},$$

$$F := \begin{bmatrix} F_1 \\ \vdots \\ F_\nu \end{bmatrix}, y_{\text{ref}}(t) := \begin{bmatrix} y_1^{\text{ref}}(t) \\ \vdots \\ y_\nu^{\text{ref}}(t) \end{bmatrix}, u(t) := \begin{bmatrix} u_1(t) \\ \vdots \\ u_\nu(t) \end{bmatrix},$$

$$e(t) := \begin{bmatrix} e_1(t) \\ \vdots \\ e_\nu(t) \end{bmatrix}, z(t) := \begin{bmatrix} z_1(t) \\ \vdots \\ z_\nu(t) \end{bmatrix}.$$

6.3 Optimal Control Parameters

It is desired to find the parameters of the controller (6.3) to minimize the quadratic performance index (6.2). Define:

$$\overline{A} := \begin{bmatrix} A + BKC_{\text{m}} & BQ \\ RC & S \end{bmatrix}, \bar{C} := \begin{bmatrix} C \; 0 \end{bmatrix}, \overline{C}_u := \begin{bmatrix} KC_{\text{m}} \; Q \end{bmatrix}.$$

Assume for now that the reference input is zero. This implies that $e(t) = y(t)$. Assume also that $\breve{Q} = I_r$ and $\breve{R} = \varepsilon I_m$ in (6.2). We will have:

$$J = \int_0^\infty \big(y'(t)y(t) + \varepsilon u'(t)u(t)\big)\mathrm{d}t = \begin{bmatrix} x'(0) \; z'(0) \end{bmatrix} \Gamma \begin{bmatrix} x(0) \\ z(0) \end{bmatrix},$$

where $\Gamma > 0$ is a solution of the following Lyapunov equation:

$$\bar{A}'\Gamma + \Gamma\bar{A} = -\big(\bar{C}'\bar{C} + \varepsilon\,\bar{C}_u'\bar{C}_u\big). \tag{6.5}$$

Since the initial state of the controller is assumed to be zero, the performance index can be simplified to:

$$J = x'(0)\begin{bmatrix} I & 0 \end{bmatrix} \Gamma \begin{bmatrix} I \\ 0 \end{bmatrix} x(0).$$

The optimal decentralized output feedback control parameters depend on the initial conditions of the plant. This means that we cannot find unique optimal control parameters for all initial states. Thus, we will use a modified version of the quadratic performance index as follows:

$$J = \mathrm{E}\left\{ \int_0^\infty \big(y'(t)y(t) + \varepsilon u'(t)u(t)\big)\mathrm{d}t \right\},$$

where E{.} denotes the expectation operator [1]. Consider now the following set of initial plant states:

$$x_1(0) = \begin{bmatrix} 1 \\ 0 \\ \vdots \\ 0 \end{bmatrix}, x_2(0) = \begin{bmatrix} 0 \\ 1 \\ \vdots \\ 0 \end{bmatrix}, \ldots, x_n(0) = \begin{bmatrix} 0 \\ 0 \\ \vdots \\ 1 \end{bmatrix},$$

which are uniformly distributed on a unit sphere, span the whole space and are orthonormal, i.e. $x_i'(0)x_j(0) = \begin{cases} 1 & i = j \\ 0 & i \neq j \end{cases}$ for any $i, j = 1, \ldots, n$. Then,

$$\begin{aligned} J &= \frac{1}{n}\left(x_1'(0)\begin{bmatrix} I & 0 \end{bmatrix} \Gamma \begin{bmatrix} I \\ 0 \end{bmatrix} x_1(0) + \cdots + x_n'(0)\begin{bmatrix} I & 0 \end{bmatrix} \Gamma \begin{bmatrix} I \\ 0 \end{bmatrix} x_n(0) \right) \\ &= \frac{1}{n}\mathrm{trace}\left(\begin{bmatrix} I & 0 \end{bmatrix} \Gamma \begin{bmatrix} I \\ 0 \end{bmatrix} \right), \end{aligned}$$

where trace(.) denotes the sum of the diagonal elements of any matrix. Therefore, we will have the following parameter optimization problem:

$$\min_{S,R,Q,K} \ \mathrm{trace}\left(\begin{bmatrix} I & 0 \end{bmatrix} \Gamma \begin{bmatrix} I \\ 0 \end{bmatrix} \right),$$

subject to the constraint:

$$\mathrm{Re}\big(\mathrm{sp}(\overline{A})\big) < 0 \quad \text{(closed-loop stability constraint)}.$$

Suppose now that the reference and disturbance inputs are constant signals. Then, the following controller can solve the decentralized robust servomechanism problem:

$$\dot{z}_i(t) = e_i(t)$$
$$u_i(t) = Q_i z_i(t) + K_i y_i^{\mathrm{m}}(t), \quad i = 1, \ldots, \nu.$$

Note that this is, in fact, an integrator whose input is the error signal and whose output is a signal, which, together with the measured output, generates the control input. The augmented system in this case can be written as follows:

$$\begin{bmatrix} \dot{x}(t) \\ \dot{z}(t) \end{bmatrix} = \begin{bmatrix} A + BKC_{\mathrm{m}} & BQI \\ C & 0 \end{bmatrix} \begin{bmatrix} x(t) \\ z(t) \end{bmatrix} + \begin{bmatrix} 0 \\ -I \end{bmatrix} y_{\mathrm{ref}}(t) + \begin{bmatrix} E \\ F \end{bmatrix} \omega(t),$$
$$y(t) = \begin{bmatrix} C & 0 \end{bmatrix} \begin{bmatrix} x(t) \\ z(t) \end{bmatrix},$$
$$u(t) = \begin{bmatrix} KC_{\mathrm{m}} & QI \end{bmatrix} \begin{bmatrix} x(t) \\ z(t) \end{bmatrix}.$$

6.4 Optimal Decentralized Control in the Context of Servomechanism Problem

In practice, it is more desirable to use the following performance index (in the context of robust servomechanism problem with constant reference and disturbance signals):

$$J = \mathrm{E}\left\{ \int_0^\infty \left(z'(t)z(t) + \varepsilon u'(t)u(t) \right) \mathrm{d}t \right\}.$$

This performance index uses the integral of the error instead of the error itself. In this case, we will have:

$$z(t) = \begin{bmatrix} 0 & I \end{bmatrix} \begin{bmatrix} x(t) \\ z(t) \end{bmatrix},$$
$$u(t) = \begin{bmatrix} KC_{\mathrm{m}} & QI \end{bmatrix} \begin{bmatrix} x(t) \\ z(t) \end{bmatrix},$$

and the matrix $\bar{C}$ in (6.5) will be given by:

$$\bar{C} = \begin{bmatrix} 0_{n \times n} & I_{n \times (r \,.\, q)} \end{bmatrix}.$$

One can use sequential optimization given in [1] to solve this problem. MATLAB's *fmincon* (*fminsearch* and *fmins* in older versions of MATLAB) is a multidimensional unconstrained nonlinear minimization function which uses the Nelder–Mead simplex (direct search) method and can also be employed here to find the optimal control parameters. The general format for this MATLAB function is:

G_opt = fmincon('perfIndex', G_0, options, p)

where *perfIndex* denotes the name of the objective function which we need to create to compute the performance index for *p* (any given set of arguments passed to the objective function) and G_0 (initial values for the parameters to be optimized). The parameters in *options* include the maximum number of iterations, termination tolerance, etc., and can be created by using the *optimset* command.

Remark 6.1 It is to be noted that if the initial parameters G_0 do not stabilize the system, the algorithm will not converge in most of the cases but if the initial parameters stabilize the system, it will certainly converge to the optimal solution.

Remark 6.2: One can add an *if* statement at the beginning of the objective function *perfIndex* to impose any given constraint by defining a penalty function (the most important constraint here is the stability condition that can be verified from the real part of the modes of the closed-loop system).

6.5 Summary

In this chapter, the optimal decentralized control problem is studied, where it is desired to design a decentralized controller which minimizes a prescribed quadratic function of the input energy and tracking error. The performance index is minimized over a set of uniformly distributed initial conditions on a sphere. For the case when a servomechanism problem is to be solved for constant signals while minimizing the quadratic performance index, a slightly different problem formulation is provided. It is shown that in both cases, the optimal control parameters can be obtained by solving a Lyapunov equation.

6.6 Problems

Hint: In the next four problems, a controller of the following form is to be used:

$$\begin{aligned} \dot{z}_i(t) &= S_i z_i(t) + R_i e_i(t) \\ u_i(t) &= Q_i z_i(t) + K_i y_i^{\mathrm{m}}(t), \quad i = 1, 2, 3, 4 \end{aligned}$$

where

$$S = \mathrm{zeros}(4), \quad R = \mathrm{eye}(4)$$

Also, the "fmincon" function on MATLAB is to be used with the following expression as the performance index:

$$\operatorname{trace}\big(\operatorname{lyap}\big(\bar{A}', \big(\bar{C}'\bar{C} + \varepsilon \bar{C}_u' \bar{C}_u\big)\big)\big)$$

where

$$\bar{C} = \begin{bmatrix} 0_{8\times 8} & I_{4\times 4} \end{bmatrix}$$

and a penalty function equal to $J = 10^9$ for any $\bar{A}$ with the right-half plane eigenvalues during the optimization process.

1. Consider the 4-input, 4-output furnace problem (given in furnace.mat). It is desired to find a controller to solve the robust servomechanism problem for this system for the case of constant reference signals and constant disturbance signals. Design a centralized controller which accomplishes this, where the controller gain matrices are forced so as to minimize the following performance index:

$$J = \mathrm{E}\left\{ \int_0^\infty \big(\Delta z'(t)\Delta z(t) + \varepsilon \Delta u'(t)\Delta u(t)\big)\mathrm{d}t \right\}, \quad \varepsilon = 10^{-5}$$

with $\Delta z(t)$ and $\Delta u(t)$ representing deviations from the equilibrium point $\Delta z(t) = z(t) - z_{\mathrm{eq}}$, $\Delta u(t) = u(t) - u_{eq}$, where $z(t)$ is the output of the servo-compensator, and $u(t)$ is the control input.

   (a) Determine the eigenvalues of the resultant closed-loop system.
   (b) Plot the response of the closed-loop system for the case of $y_{\mathrm{ref}}(t)$ equal to $\begin{bmatrix} 1(t) \\ 0 \\ 0 \\ 0 \end{bmatrix}, \begin{bmatrix} 0 \\ 1(t) \\ 0 \\ 0 \end{bmatrix}, \begin{bmatrix} 0 \\ 0 \\ 1(t) \\ 0 \end{bmatrix}, \begin{bmatrix} 0 \\ 0 \\ 0 \\ 1(t) \end{bmatrix}$ and zero initial conditions, where $1(t)$ represents the unit-step signal.

2. Repeat Problem 1, using the following decentralized controller:

$$u(t) = \Theta y(t) + \Phi \int_0^\infty e(t)\mathrm{d}t, \quad e(t) = y(t) - y_{\mathrm{ref}}(t)$$

where:

$$\Theta = \begin{bmatrix} \theta_1 & 0 & 0 & 0 \\ 0 & \theta_2 & 0 & 0 \\ 0 & 0 & \theta_3 & 0 \\ 0 & 0 & 0 & \theta_4 \end{bmatrix}, \quad \Phi = \begin{bmatrix} \phi_1 & 0 & 0 & 0 \\ 0 & \phi_2 & 0 & 0 \\ 0 & 0 & \phi_3 & 0 \\ 0 & 0 & 0 & \phi_4 \end{bmatrix}.$$

Hint: Use $\theta_i = -1,\ \phi_i = -1, i = 1, 2, 3, 4$ *as the starting point.*

3. Repeat Problem 1, using the following decentralized controller:

$$u(t) = \theta_1 I y(t) + \phi_1 I \int_0^\infty e(t)\mathrm{d}t,$$

where $\theta_1,\ \phi_1$ are scalars. Compare the performance of this controller with that obtained in Problems 1 and 2.

4. Repeat Problem 3 using $\varepsilon = 10^{-10}$ in the system's performance index after imposing the following constraints:

$$|\theta_1| < 500, \quad |\phi_1| < 500$$

5. Consider the following system:

$$\begin{aligned} \dot{x}(t) &= Ax(t) + \sum_{i=1}^{\nu} B_i u_i(t) + E\omega(t) \\ y_i(t) &= C_i x(t) + D_i u_i(t) + F_i \omega(t), \quad i = 1, \ldots, \nu \\ y_i^{\mathrm{m}}(t) &= C_i^{\mathrm{m}} x(t) + D_i^{\mathrm{m}} u_i(t) + F_i^{\mathrm{m}} \omega(t), \quad i = 1, \ldots, \nu \end{aligned}$$

where $x(t) \in R^n$ is the state vector, and $u_i(t) \in R^{m_i}$ and $y_i(t) \in R^{r_i}$ are the input and output to be regulated, respectively, of the ith local control station ($i = 1, \ldots, \nu$); $y_i^{\mathrm{m}}(t) \in R^{r_i^{\mathrm{m}}}$ ($i = 1, \ldots, \nu$) is an output which is available for measurement at the ith control station, and $\omega(t) \in R^{\Omega}$ is a disturbance. Furthermore, $A \in R^{n \times n}$, $B_i \in R^{n \times m_i}$, $C_i \in R^{r_i \times n}$, $D_i \in R^{r_i \times m_j}$, $F_i \in R^{r_i \times \Omega}$, $C_i^{\mathrm{m}} \in R^{r_i^{\mathrm{m}} \times n}$, $D_i^{\mathrm{m}} \in R^{r_i^{\mathrm{m}} \times m_j}$ and $F_i^{\mathrm{m}} \in R^{r_i^{\mathrm{m}} \times \Omega}$ ($i = 1, \ldots, \nu$) are constant matrices. Let the reference input for the ith local control station be zero, which implies that the error $e_i(t)$ in the ith local control station is equal to $y_i(t)$.Consider the following decentralized controller for the system:

$$\begin{aligned} \dot{z}_i(t) &= S_i z_i(t) + R_i e_i(t) \\ u_i(t) &= Q_i z_i(t) + K_i y_i^{\mathrm{m}}(t), \quad i = 1, \ldots, \nu \end{aligned}$$

where the initial conditions of the controller states are assumed to be zero $z_i(0) = 0,\ i = 1, \ldots, \nu$. It is desired to minimize the quadratic performance index:

$$J = \mathrm{E}\left\{ \int_{0}^{\infty} \left(y'(t)y(t) + \varepsilon u'(t)u(t) \right) \mathrm{d}t \right\}.$$

Given the following set of initial plant states:

$$x_1(0) = \begin{bmatrix} \theta_1 \\ 0 \\ \vdots \\ 0 \end{bmatrix}, x_2(0) = \begin{bmatrix} 0 \\ \theta_2 \\ \vdots \\ 0 \end{bmatrix}, \ldots, x_n(0) = \begin{bmatrix} 0 \\ 0 \\ \vdots \\ \theta_n \end{bmatrix},$$

where θ_i, $i = 1, \ldots, \nu$, are known scalars. Assume that the solution of the Lyapunov equation $\bar{A}'\Gamma + \Gamma\bar{A} = -(\bar{C}'\bar{C} + \varepsilon\bar{C}_u'\bar{C}_u)$, where

$$\bar{A} := \begin{bmatrix} A + BKC_{\mathrm{m}} & BQ \\ RC & S \end{bmatrix}, \bar{C} := \begin{bmatrix} C & 0 \end{bmatrix}, \bar{C}_u := \begin{bmatrix} KC_{\mathrm{m}} & Q \end{bmatrix},$$

results in the following $(n + rq) \times (n + rq)$ diagonal matrix:

$$\Gamma = \mathrm{diag}\left(\begin{bmatrix} 1 \ 2 \ \cdots \ n \ n+1 \ \cdots \ n+rq \end{bmatrix} \right).$$

Find a closed-form formula for the performance index J.

6. Consider the linearized model of Problem 3.3 with $g = 10$, $l = 10$, $m = 0.01$, $k = 0.02$, $a = 5$.

   (a) It is desired to find the decentralized controller parameters to solve the decentralized robust servomechanism problem and minimize the performance index:

   $$J = \mathrm{E}\left\{ \int_{0}^{\infty} \left(z'(t)z(t) + \varepsilon u'(t)u(t) \right) \mathrm{d}t \right\},$$

   where the expected value is taken over the following initial conditions:

   $$\begin{bmatrix} x(0) \\ z_1(0) \\ z_2(0) \end{bmatrix} = \begin{bmatrix} 1 \\ 0 \\ 0 \\ 0 \\ 0 \\ 0 \end{bmatrix}, \begin{bmatrix} 0 \\ 1 \\ 0 \\ 0 \\ 0 \\ 0 \end{bmatrix}, \begin{bmatrix} 0 \\ 0 \\ 1 \\ 0 \\ 0 \\ 0 \end{bmatrix}, \begin{bmatrix} 0 \\ 0 \\ 0 \\ 1 \\ 0 \\ 0 \end{bmatrix}, \begin{bmatrix} 0 \\ 0 \\ 0 \\ 0 \\ 1 \\ 0 \end{bmatrix}, \begin{bmatrix} 0 \\ 0 \\ 0 \\ 0 \\ 0 \\ 1 \end{bmatrix}.$$

   Find the controller parameters and the corresponding minimum performance index for $\varepsilon = 0.1$ and $\varepsilon = 0.001$.

(b) Show the simulation results of the optimal controller obtained for $\varepsilon = 0.1$. In your simulations, use the linearized model with the initial condition:

$$\begin{bmatrix} x(0) \\ z_1(0) \\ z_2(0) \end{bmatrix} = \begin{bmatrix} 1 \\ 1 \\ 1 \\ 1 \\ 1 \\ 1 \end{bmatrix},$$

and the disturbance signal $\omega(t) = 0.1 \times 1(t)$, and the reference signal $\begin{bmatrix} y_1^{\text{ref}}(t) \\ y_2^{\text{ref}}(t) \end{bmatrix} = \begin{bmatrix} 0.1 \times 1(t) \\ 0 \end{bmatrix}$, where, similar to Problem 6.1, $1(t)$ represents the unit-step function.

Reference

1. W.S. Levine, M. Athans, On the determination of the optimal constant output feedback gains for linear multivariable systems. IEEE Trans. Autom. Control **AC-15**, 44–48 (1970)

Chapter 7
Optimal Performance in the Decentralized Setting

In Chap. 5, we demonstrated that NLTV controllers have an advantage over LTI controllers in stabilizing an LTI decentralized system. In Chap. 6, we explored the design of optimal LTI controllers. In this chapter, we explore a class of problems for which the optimal controller is not normally LTI; in this context, the goal is to obtain "near-optimal" performance, in the sense that we can get as close to the optimal level of performance as desired.

7.1 Introduction

In Chap. 5, we introduced the notion of a quotient system and its DFMs, denoted QFMs. In fact, if the system has no unstable QFMs, we showed that we can always construct a linear periodic stabilizing controller, and several controller design procedures are provided. In this chapter, we turn to the much more demanding goal of designing a controller which provides optimal (or near optimal) performance, with a focus on situations in which the optimal controller is not LTI. It turns out that the connectedness of the system graph plays an important role.

An outline of the chapter is as follows:

(i) In Sect. 7.2, we provide a formal setup of the control problem and the measure of performance.
(ii) In Sect. 7.3, we provide an example demonstrating the gap between the optimal centralized performance and the optimal decentralized performance. Indeed, we show that the gap can be arbitrarily large.

E. J. Davison et al., *Decentralized Control of Large-Scale Systems*,
https://doi.org/10.1007/978-1-4419-6014-6_7

(iii) In Sects. 7.4 and 7.5, we consider a class of systems for which the LQR-optimal decentralized performance equals that of the optimal centralized performance; indeed, this class of systems includes some which cannot even be stabilized using an LTI controller. In Sect. 7.4, we consider the simplest decentralized case in which there are two scalar channels, while in Sect. 7.5, we explain how to remove these restrictions.

7.2 The Plant and the Control Objective

Here we consider the strictly proper plant

$$\left.\begin{aligned} \dot{x} &= Ax + \textstyle\sum_{i=1}^{\nu} B_i u_i, \quad x(t_0) = x_0 \\ y_i &= C_i x, \quad i = 1, \ldots, \nu \end{aligned}\right\} \tag{7.1}$$

with $x(t) \in \mathbf{R}^n$, $u_i(t) \in \mathbf{R}^{m_i}$ and $y_i(t) \in \mathbf{R}^{r_i}$ for $i = 1, \ldots, \nu$; we set $m = \sum_{i=1}^{\nu} m_i$ and $r = \sum_{i=1}^{\nu} r_i$. Associated with this model are the quantities associated with the centralized version of the plant:

$$B := \begin{bmatrix} B_1 & \cdots & B_\nu \end{bmatrix},$$

$$y = \begin{bmatrix} y_1 \\ \vdots \\ y_\nu \end{bmatrix}, \quad u = \begin{bmatrix} u_1 \\ \vdots \\ u_\nu \end{bmatrix}, \quad C := \begin{bmatrix} C_1 \\ \vdots \\ C_\nu \end{bmatrix}, \quad \mathcal{O}(C_i, A) := \begin{bmatrix} C_i \\ C_i A \\ \vdots \\ C_i A^{n-1} \end{bmatrix}.$$

We **assume** that the system is centrally controllable and observable, i.e. (A, B) is controllable, and (C, A) is observable.

The goal here is to design a linear time-varying controller which not only provides closed-loop stability but also provides near-optimal performance, with performance measured in an LQR framework. So fix positive definite symmetric matrices $Q \in \mathbf{R}^{n\times n}$ and $R \in \mathbf{R}^{m\times m}$ and consider the quadratic performance index

$$\int_0^\infty [x^T(t)Qx(t) + u^T(t)Ru(t)]dt. \tag{7.2}$$

With $t_0 = 0$ and $x_0 \in \mathbf{R}^n$, the LQR problem is to find the controller which minimizes this cost; the problem can be posed in both the centralized and decentralized setting. As is well-known, the **centralized LQR-optimal controller** is state feedback:

$$u = Fx,$$

which gives rise to an optimal cost (measured via (7.2)) of the form $J^{opt}(x_0) = x_0^T P x_0$, with P the positive definite solution of the associated Riccati equation

$$A^T P + PA - PB\,R^{-1}B^T P + Q = 0.$$

On the other hand, in the decentralized setting the problem is much harder and no fully general synthesis methods exist. In the next section, we consider an example for which there is a large gap between the optimal centralized and optimal decentralized performance.

7.3 The Performance Gap Between the Centralized and Decentralized Cases

In this section,[1] we show that the gap between the optimal centralized performance and the optimal decentralized performance can be arbitrarily large. We do so by considering a very simple second order example, carefully chosen so that the optimal centralized controller is highly non-diagonal, which results in the performance gap. The plant that we consider is parameterized by $k > 0$ and described by

$$\dot{x} = \underbrace{\begin{bmatrix} 0 & 0 \\ 0 & 0 \end{bmatrix}}_{=:A} x + \underbrace{\begin{bmatrix} 1 \\ 0 \end{bmatrix}}_{=:B_1} u_1 + \underbrace{\begin{bmatrix} 0 \\ 1 \end{bmatrix}}_{=:B_2} u_2, \; x(0) = x_0 \tag{7.3}$$

$$y_1 = \underbrace{\begin{bmatrix} k^{1/2} & k \end{bmatrix}}_{=:C_1(k)} x \tag{7.4}$$

$$y_2 = \underbrace{\begin{bmatrix} 0 & k^{1/4} \end{bmatrix}}_{=:C_2(k)} x, \tag{7.5}$$

with $x(t) \in \mathbf{R}^2$, $u_i(t) \in \mathbf{R}$ and $y_i(t) \in \mathbf{R}$ for $i = 1, 2$. This choice is a much simplified version of the example presented in [2].

It is easy to see that this system is centrally controllable and observable for every $k > 0$. Furthermore, it is easy to check that the system has no decentralized fixed modes, so there exists a stabilizing decentralized LTI controller. The associated plant transfer function is

$$\begin{bmatrix} Y_1(s) \\ Y_2(s) \end{bmatrix} = \underbrace{\begin{bmatrix} \frac{k^{1/2}}{s} & \frac{k}{s} \\ 0 & \frac{k^{1/4}}{s} \end{bmatrix}}_{=:G(s)} \begin{bmatrix} U_1(s) \\ U_2(s) \end{bmatrix}.$$

[1] This section is derived, in part, from [2].

It is clear that the associated graph is not strongly connected, a fact which turns out to be important.

Here, we consider the very natural quadratic performance index of

$$\int_0^\infty [y(t)^T y(t) + u(t)^T u(t)]dt; \tag{7.6}$$

in the notation of Sect. 7.2 this means that $R(k) = I$ and $Q(k) := C(k)^T C(k)$, which are positive definite matrices, so this fits the standard paradigm of this chapter. The LQR problem is to find, for a given $x_0 \in \mathbf{R}^n$, the control signal u which minimizes this cost. As mentioned in Sect. 7.2, the optimal centralized controller is state feedback of the form

$$u = F(k)x; \tag{7.7}$$

since $C(k)$ is invertible, we can also view this as static output feedback of the form

$$u = F(k)C(k)^{-1}y.$$

We can obtain $F(k)$ by solving an associated algebraic Riccati equation (ARE) for the unique positive definite symmetric solution $P(k)$; in this case, the equation is particularly simple:

$$-P(k)^2 + C(k)^T C(k) = 0. \tag{7.8}$$

Once we have the solution $P(k)$, the feedback gain is given by

$$F(k) = -B^T P(k) = -P(k)$$

and the optimal cost, for a given initial condition x_0, is exactly

$$J_k^{opt}(x_0) = x_0^T P(k) x_0.$$

We will now prove, for the carefully chosen initial condition of $x_0 = \begin{bmatrix} 1 \\ 0 \end{bmatrix}$, that the performance provided by any decentralized controller will be significantly larger than that provided by the optimal (centralized) controller, with the gap increasing as $k \to \infty$.

To proceed, we first examine $J_k^{opt}(x_0)$ in more detail. Since it is hard to get a closed-form description of it, instead we will opt for a crude upper bound for the initial condition of interest.

Proposition 1 *For large $k > 0$, we have* $J_k^{opt}(\begin{bmatrix} 1 \\ 0 \end{bmatrix}) \leq 2k^{1/4}$.

Proof To proceed, we need to analyse the ARE (7.8). Let us write the symmetric solution $P(k)$ as

$$P(k) = \begin{bmatrix} p_1(k) & p_2(k) \\ p_2(k) & p_3(k) \end{bmatrix};$$

henceforth, we will typically suppress the argument for clarity. Hence, (7.8) can be rewritten as

$$\begin{bmatrix} p_1^2 + p_2^2 & p_2(p_1 + p_3) \\ p_2(p_1 + p_3) & p_2^2 + p_3^2 \end{bmatrix} = \begin{bmatrix} k & k^{3/2} \\ k^{3/2} & k^2 + k^{1/2} \end{bmatrix};$$

because of the symmetry, we end up with three equations and three unknowns:

$$p_1^2 + p_2^2 = k \tag{7.9}$$

$$p_2(p_1 + p_3) = k^{3/2} \tag{7.10}$$

$$p_2^2 + p_3^2 = k^2 + k^{1/2}. \tag{7.11}$$

The quantity of interest is exactly

$$J_k^{opt}(\begin{bmatrix} 1 \\ 0 \end{bmatrix}) = p_1(k).$$

To proceed, the fact that P is positive definite means that its diagonal elements p_1 and p_3 are both positive. From (7.9), we see that

$$p_2^2 \in [0, k],$$

so from (7.11), we have that

$$p_3^2 = k^2 + k^{1/2} - p_2^2 \leq k^2 + k^{1/2},$$

which implies that

$$p_3 \leq (k^2 + k^{1/2})^{1/2}. \tag{7.12}$$

Also, from the fact that $p_1 > 0$ and (7.9), we see that

$$p_1 \in (0, k^{1/2}];$$

if we substitute this and (7.12) into (7.10), we see that

$$\begin{aligned} p_2^2 &= \frac{k^3}{(p_1+p_3)^2} \\ &\geq \frac{k^3}{[k^{1/2}+(k^2+k^{1/2})^{1/2}]^2} \\ &= \frac{k^3}{k+k^2+k^{1/2}+2k^{1/2}(k^2+k^{1/2})^{1/2}}. \end{aligned}$$

If we now use this lower bound on p_2^2 in (7.9), we end up with

$$\begin{aligned} p_1^2 &= k - p_2^2 \\ &\leq k - \frac{k^3}{k+k^2+k^{1/2}+2k^{1/2}(k^2+k^{1/2})^{1/2}} \\ &= \frac{k^2+k^{3/2}+2k^{3/2}(k^2+k^{1/2})^{1/2}}{k+k^2+k^{1/2}+2k^{1/2}(k^2+k^{1/2})^{1/2}}; \end{aligned}$$

for large $k > 0$, we have that

$$p_1^2 \leq 4k^{1/2},$$

so

$$p_1 \in [0, 2k^{1/4}].$$

Hence, for large k, we have

$$J_k^{opt}(\begin{bmatrix} 1 \\ 0 \end{bmatrix}) \leq 2k^{1/4},$$

as desired.

□

Now we turn to the problem of obtaining a lower bound on the optimal decentralized performance. To proceed, we need to identify the admissible class of controllers. To prove a general result, we wish to impose as few constraints as possible. To this end, we allow our controller to be input–output of the form

$$u_1 = K_1(k)y_1, \tag{7.13}$$

and

$$u_2 = K_2(k)y_2, \tag{7.14}$$

where each $K_i(k)$ is a general map from the set of piecewise continuous signals to itself. First we impose some very natural assumptions about the well-posedness of the closed-loop system. Before stating this, notice that in our example the information flow is upwards from channel two to channel one, so it is reasonable that the second subsystem be well-posed in its own right.

Well-Posedness Assumption:

(i) The whole system is well-posed in the sense that, for every $x_0 \in \mathbf{R}^2$, there exist unique piecewise continuous signals u_1, u_2, y_1, y_2 and x which satisfy (7.3)–(7.5), (7.13) and (7.14).
(ii) The second subsystem is well-posed in the sense that, when (7.14) is applied to the plant subsystem

$$\dot{x}_2 = u_2, \quad x_2(0) = x_{2_0}, \tag{7.15}$$

$$y_2 = k^{1/4} x_2, \tag{7.16}$$

then for every $x_{2_0} \in \mathbf{R}$, we have that there exist unique piecewise continuous signals u_2, y_2 and x_2 which satisfy (7.14), (7.15) and (7.16).

Given a decentralized controller (7.13) and (7.14) which satisfies the Well-Posedness Assumption, for a given $x_0 \in \mathbf{R}^n$ let $J_k(x_0)$ denote the corresponding cost. A natural way to measure the relative performance is via

$$\gamma_k := \sup_{x_0 \neq 0} \frac{J_k(x_0)}{J_k^{opt}(x_0)}.$$

Of course, we do not want to rule out the case of $x_0 = 0$; for that initial condition, we have $J_k^{opt}(x_0) = 0$, so it is reasonable to require this in the decentralized setting as well. This leads to the

Performance Regularization Assumption: $J_k(0) = 0$.

This brings us to the main result of this section.

Theorem 1 *If the controller (7.13) and (7.14) is applied to the plant (7.3)–(7.5) and both the Well-Posedness Assumption and the Performance Regularization Assumption hold, then*

$$\lim_{k \to \infty} \gamma_k = \infty.$$

Proof Fix $k > 0$ and suppose that the controller (7.13) and (7.14) is applied to the plant (7.3)–(7.5) and both the Well-Posedness Assumption and the Performance Regularization Assumption hold. To proceed, we consider two different initial conditions.

First let us consider an initial condition of zero. By the Performance Regularization Assumption, it must be that the performance index is zero, so the corresponding state and control signals must be zero almost everywhere.[2] In particular, the second control signal, which we label u_2^0 (to emphasize that it corresponds to the case of a zero initial condition), is zero almost everywhere.

Now let us consider an initial condition of $x_0 = \begin{bmatrix} 1 \\ 0 \end{bmatrix}$; let x_1, x_2, y_1, y_2, u_1 and u_2 denote the corresponding closed-loop signals. Since both this case and the previous case have an initial condition on x_2 of zero, by the second part of the Well-Posedness Assumption it must be that $u_2 = u_2^0$, which means, in particular, that u_2 is zero almost everywhere. Solving the second differential equation of the plant, we conclude that $y_2(t) = 0$ for $t \geq 0$, so the first differential equation of the plant simplifies to

$$\dot{x}_1 = u_1, \quad x_1(0) = 1,$$

$$y_1 = k^{1/2} x_1,$$

and the cost becomes

$$J_k(\begin{bmatrix} 1 \\ 0 \end{bmatrix}) = \int_0^\infty [y_1^2(t) + u_1^2(t)]\, dt = \int_0^\infty [k x_1(t)^2 + u_1^2(t)]\, dt.$$

But a lower bound on the RHS is well-known: it arises from a first order LQR problem, and with p_k representing the positive definite solution of the first-order quadratic equation

$$p_k^2 - k = 0,$$

namely $p_k = k^{1/2}$, we have that

$$J_k(\begin{bmatrix} 1 \\ 0 \end{bmatrix}) \geq p_k = k^{1/2}.$$

From Proposition 1, we know that for large k, we have

$$J_k^{opt}(\begin{bmatrix} 1 \\ 0 \end{bmatrix}) \leq 2k^{1/4}.$$

---

[2]We say that a signal is zero almost everywhere if the set of times where this control signal is nonzero is of Lebesgue measure zero.

By the definition of γ_k, it follows that for large k

$$\gamma_k = \sup_{x_0 \neq 0} \frac{J_k(x_0)}{J_k^{opt}(x_0)} \geq \frac{J_k(\begin{bmatrix} 1 \\ 0 \end{bmatrix})}{J_k^{opt}(\begin{bmatrix} 1 \\ 0 \end{bmatrix})} \geq \frac{k^{1/2}}{2k^{1/4}} = \frac{1}{2}k^{1/4}.$$

So we see that $\lim_{k\to\infty} \gamma_k = \infty$, as desired.

□

Hence, we conclude that, under some reasonable regularity and well-posedness assumptions, there can be an arbitrarily large gap between the LQR-optimal centralized performance and the LQR-optimal decentralized performance. Quite surprisingly, it turns out that this gap can only arise when the associated graph of the system is not strongly connected. In the next two sections, we will prove that no such gap exists for the strongly connected case, even when the system has unstable DFMs.

7.4 The Strongly Connected Case with Two Scalar Channels

In this section[3] the goal, under suitable assumptions, is to construct an LQR "near-optimal" decentralized controller, not necessarily LTI. As discussed in Chap. 5, if the plant is centrally controllable and observable and the associated system graph is strongly connected, then there always exists an LPC which stabilizes it, even in the presence of unstable DFMs (because of these assumptions, it has no QFMs). It turns out, in this situation, that we can actually obtain a level of LQR performance as close as desired to the optimal. Since the key ideas are present in the two scalar channel case, that is what we will focus on in this section; in Sect. 7.5, we will show how to easily extend this to the general case. So here we assume that

(i) There are two channels, i.e. $\nu = 2$.
(ii) The channels are scalar, i.e., $m_1 = r_1 = 1$ and $m_2 = r_2 = 1$.
(iii) The associated graph is strongly connected, which means, in particular, that the transfer functions $C_1(sI - A)^{-1}B_2$ and $C_2(sI - A)^{-1}B_1$ are both nonzero.

The approach uses a linear periodic sampled-data controller, although it is more complicated that those of Sects. 5.5 and 5.6, which use a generalized hold and generalized sampler, respectively, but it is of comparable complexity to that of the approach of Sect. 5.7, which we did not cover in complete detail. The idea here is to carry out probing, estimation and control in the following way. We first construct the optimal centralized state feedback controller offline. Next, we construct a periodic controller: while at all times applying the current estimate of the optimal control signal, during

[3] This section is derived, in part, from [3].

the first part of each period we probe the system so that we can estimate a sufficiently large part of the state in each channel so that, if the estimates are combined, then we have an estimate of the full state. During the second part of the period, we probe the system once again to pass the state information in channel one to channel two and the state information in channel two to channel one; at the end of the period, an estimate of the plant state is available in each channel, so the estimate of the optimal control signal can be updated accordingly and used during the following period. The controller is parameterized by two quantities: the period $T > 0$ and the probing signal weight $\rho > 0$; we can adjust them to obtain closed-loop stability as well as closed-loop performance as close to the optimal as desired.

At this point, we freeze Q and R positive definite and symmetric; following Sect. 7.2, the optimal (centralized) control law is state feedback and of the form

$$u = Fx = \begin{bmatrix} F_1 \\ F_2 \end{bmatrix} x,$$

and the optimal cost, for a given initial condition x_0 (and with $t_0 = 0$) is of the form

$$J^{opt}(x_0) = x_0^T P x_0;$$

we label the corresponding optimal state trajectory by $x^{opt}(t)$ and the optimal control signal by $u^{opt}(t)$ (clearly both quantities are implicit functions of x_0, but we omit that dependence for clarity). The goal now is to design a linear periodic decentralized controller which stabilizes the plant in a suitable sense and which yields a level of performance as close as desired to the optimal. The controller is sampled-data with a base sampling period of h; the actual period of the controller is an integer multiple of this, which we denote by T.

7.4.1 The High-Level Idea

First of all, it is clear that a sampled-data approximation of the optimal control law of the form

$$u(t) = \begin{bmatrix} u_1(t) \\ u_2(t) \end{bmatrix} = \begin{bmatrix} F_1 \\ F_2 \end{bmatrix} x(kT), \quad t \in [kT, (k+1)T)$$

will be near-optimal if T is small enough, and we would expect that the difference in cost should tend to zero as $T \to 0$. However, normally in each channel only part of x is measurable, so typically $F_i x(kT)$ cannot be constructed from $y_i(kT)$, so we will have to somehow estimate it. If we turn off the whole control signal for $t \in [t_0, t_1]$, then

$$y_i(t) = C_i e^{A(t-t_0)} x_0, \quad t \in [t_0, t_1];$$

by differentiating this j times, we can clearly obtain

$$y_i^{(j)}(t) = C_i A^j e^{A(t-t_0)} x_0, \quad t \in [t_0, t_1],$$

which, when evaluated at $t = t_0$, yields $C_i A^j x_0, j \geq 0$; of course, to reduce the effect of noisy signals, it is better to use samples of $y(t)$ on the interval to estimate these quantities, just as in Sect. 5.6. Hence, if we turn off the control input on an interval, then by measuring $y_i(t)$ only, we can easily obtain an estimate of $\mathcal{O}(C_i, A)x_0$ in channel i. If we were to combine $\mathcal{O}(C_1, A)x_0$ and $\mathcal{O}(C_2, A)x_0$, then it follows from the hypothesis that the system is centrally observable that we can compute x_0, and therefore, Fx_0; the problem is that $\mathcal{O}(C_1, A)x_0$ is only available in channel one and $\mathcal{O}(C_2, A)x_0$ is only available in channel two. The obvious solution is to pass the missing information between the channels, i.e. pass knowledge of $\mathcal{O}(C_1, A)x_0$ from channel one to channel two and knowledge of $\mathcal{O}(C_2, A)x_0$ from channel two to channel one. Of course, the two quantities $\mathcal{O}(C_1, A)x_0$ and $\mathcal{O}(C_2, A)x_0$ contains $2n$ pieces of data, so some of this is redundant. To this end, there exist n linearly independent rows of $\begin{bmatrix} \mathcal{O}(C_1, A) \\ \mathcal{O}(C_1, A) \end{bmatrix}$ - we can take $n_1 \in \{0, \ldots, n\}$ rows of $\mathcal{O}(C_1, A)$ and $n_2 \in \{0, \ldots, n\}$ rows of $\mathcal{O}(C_2, A)$ (with $n_1 + n_2 = n$), and it is easy to prove that we can take the first n_1 rows of $\mathcal{O}(C_1, A)$ and the first n_2 rows of $\mathcal{O}(C_2, A)$. To this end, we can then define

$$\begin{aligned} \xi^1(t) &:= \begin{bmatrix} I_{n_1} & 0 \end{bmatrix} \mathcal{O}(C_1, A)x(t), \\ \xi^2(t) &:= \begin{bmatrix} I_{n_2} & 0 \end{bmatrix} \mathcal{O}(C_2, A)x(t); \end{aligned}$$

with

$$\mathcal{O}_{est} := \begin{bmatrix} \begin{bmatrix} I_{n_1} & 0 \end{bmatrix} & 0 \\ 0 & \begin{bmatrix} I_{n_2} & 0 \end{bmatrix} \end{bmatrix} \begin{bmatrix} \mathcal{O}(C_1, A) \\ \mathcal{O}(C_2, A) \end{bmatrix},$$

it follows that

$$\mathcal{O}_{est}^{-1} \begin{bmatrix} \xi^1(t) \\ \xi^2(t) \end{bmatrix} = x(t).$$

So at a high level, the idea is that we first estimate $\xi^1(t_0)$ in channel one and then pass it to channel two; similarly, we estimate $\xi^2(t_0)$ in channel two and then pass it to channel one. At this point, we would have sufficient information in each channel to estimate x_0 and therefore to estimate the optimal control signal Fx_0. Of course, we would like to do so while at the same time approximately achieving our control objective, and we would like carry out the above procedure fairly often, since the state is always changing.

To make this more precise, consider the following more detailed approach. On each period $[kT, (k+1)T)$, we apply an estimate $\hat{u}^{opt}(kT)$ of $u^{opt}(kT)$, while at the same time carrying out probing to obtain a better estimate of this quantity for use during the next period. We use the fact that the off-diagonal terms in the plant transfer function are nonzero to pass information between channels: for each element of our estimate $\hat{\xi}^1(kT)$ of $\xi^1(kT)$, we use probing to pass the information from channel one channel two, and then for each element of our estimate $\hat{\xi}^2(kT)$ of $\xi^2(kT)$, we use probing to pass the information from channel two channel one; so at time $t = (k+1)T$, we have a good estimate of $x(kT)$ in each channel, which can be used to form a good estimate of $Fx(kT)$ to generate the new control signal $\hat{u}^{opt}((k+1)T)$ for use during the following period $[(k+1)T, (k+2)T)$. For this to work, we need an effective estimation procedure, which is the next topic of discussion.

7.4.2 Estimation

To understand how to carry out estimation, consider the centralized system

$$\dot{x} = \begin{bmatrix} 0 & 1 & 0 \\ 0 & 0 & 1 \\ 0 & 0 & 0 \end{bmatrix} x + Bu,$$

$$y = Cx.$$

With $\bar{u} \in \mathbf{R}$ representing the nominal control signal and $\delta \neq 0$ representing a scaling factor, suppose that we set

$$u(t) = (1+\delta)\bar{u}, \quad t \in [0, 3h);$$

then by solving the differential equation, it is easy to see that

$$y(t) = \begin{bmatrix} 1 & t & \frac{t^2}{2} & \frac{t^3}{3!} \end{bmatrix} [\begin{bmatrix} C \\ CA \\ CA^2 \\ 0 \end{bmatrix} x(0) + \begin{bmatrix} 0 \\ CB \\ CAB \\ CA^2B \end{bmatrix} (1+\delta)\bar{u}], \quad t \in [0, 3h].$$

It follows that

$$\begin{bmatrix} 1 & & & \\ & h & & \\ & & h^2/2 & \\ & & & h^3/3! \end{bmatrix}^{-1} \begin{bmatrix} 1 & 0 & 0 & 0 \\ 1 & 1 & 1 & 1 \\ 1 & 2 & 4 & 8 \\ 1 & 3 & 9 & 27 \end{bmatrix}^{-1} \begin{bmatrix} y(0) \\ y(h) \\ y(2h) \\ y(3h) \end{bmatrix} = \begin{bmatrix} C \\ CA \\ CA^2 \\ 0 \end{bmatrix} x(0)$$
$$+ \begin{bmatrix} 0 \\ CB \\ CAB \\ CA^2B \end{bmatrix} (1+\delta)\bar{u}. \tag{7.17}$$

Hence, we can use samples of $y(t)$ at $t = 0, h, 2h, 3h$ to estimate a weighted sum of $x(0)$ and $B\bar{u}$. Now set

$$u(t) = (1 - \delta)\bar{u}, \quad t \in [3h, 6h);$$

if h is small, notice that the effect of the probing signal on this interval approximately cancels out the effect of the probing on the previous interval. Furthermore, using the same argument as above, we can obtain an estimate of

$$\begin{bmatrix} C \\ CA \\ CA^2 \\ 0 \end{bmatrix} x(3h) + \begin{bmatrix} 0 \\ CB \\ CAB \\ CA^2B \end{bmatrix} (1 - \delta)\bar{u};$$

if h is small, then this is approximately equal to

$$\begin{bmatrix} C \\ CA \\ CA^2 \\ 0 \end{bmatrix} x(0) + \begin{bmatrix} 0 \\ CB \\ CAB \\ CA^2B \end{bmatrix} (1 - \delta)\bar{u}. \tag{7.18}$$

If we take the difference of the estimates (7.17) and (7.18), then we can obtain an estimate of

$$\begin{bmatrix} 0 \\ CB \\ CAB \\ CA^2B \end{bmatrix} \delta\bar{u},$$

and hence of

$$\begin{bmatrix} 0 \\ CB \\ CAB \\ CA^2B \end{bmatrix} \bar{u}, \tag{7.19}$$

which we can then subtract from our initial quantity (7.17) to obtain an estimate of

$$\begin{bmatrix} C \\ CA \\ CA^2 \\ 0 \end{bmatrix} x(0).$$

If the transfer function $C(sI - A)^{-1}B$ is not identically zero, then by the Cayley–Hamilton theorem we see that $CA^iB = 0$ for $i > 2$, so it must be that

$$\begin{bmatrix} 0 \\ CB \\ CAB \\ CA^2B \end{bmatrix} \neq 0,$$

in which case we can easily estimate $\bar{u}$ from the estimate of (7.19). It turns out that the above idea works for a more general choice of A, the state dimension n, as well as in the decentralized setting.

So let us return to our two channel decentralized setting subject to the assumptions of this section. To proceed, we need to determine how long the estimation interval $[t_0, t_0 + \bar{n}h]$ should be. If the above analysis generalizes in a natural way (which the upcoming lemma will show to be true), and if we were to probe in channel one with

$$u_1(t) = \begin{cases} (1+\delta)\bar{u}_1 & t \in [t_0, \bar{n}h), \\ (1-\delta)\bar{u}_1 & t \in [t_0 + \bar{n}h, t_0 + 2\bar{n}h), \end{cases}$$

(and with $u_2(t) = 0$ there), then by sampling the channel one output y_1, we can obtain an accurate estimate of

$$\begin{bmatrix} C_1 \\ C_1A \\ \vdots \\ C_1A^{\bar{n}} \end{bmatrix} x(t_0) \text{ and } \begin{bmatrix} 0 \\ C_1B_1 \\ \vdots \\ C_1A^{\bar{n}-1}\bar{B}_1 \end{bmatrix} \bar{u}_1, \qquad (7.20)$$

and by sampling the channel two output y_2 we can obtain an accurate estimate of

$$\begin{bmatrix} C_2 \\ C_2A \\ \vdots \\ C_2A^{\bar{n}} \end{bmatrix} x(t_0) \text{ and } \begin{bmatrix} 0 \\ C_2B_1 \\ \vdots \\ C_2A^{\bar{n}-1}\bar{B}_1 \end{bmatrix} \bar{u}_1. \qquad (7.21)$$

So to implement our estimation and probing idea:

- In channel one, we wish to estimate $\xi^1(t_0) = \begin{bmatrix} I_{n_1} & 0 \end{bmatrix} \mathcal{O}(C_1, A)x(t_0)$, so from (7.20) we will need $\bar{n} \geq n_1 - 1$.
- In channel two, we wish to estimate $\xi^2(t_0) = \begin{bmatrix} I_{n_2} & 0 \end{bmatrix} \mathcal{O}(C_2, A)x(t_0)$, so from (7.21) we will need $\bar{n} \geq n_2 - 1$.

- In channel two, we wish to estimate $\bar{u}_1$, so from (7.21) we need $\begin{bmatrix} 0 \\ C_2B_1 \\ \vdots \\ C_2A^{\bar{n}-1}\bar{B}_1 \end{bmatrix}$ to be nonzero; hence, we need $\bar{n}$ to be at least as large as the relative degree of $C_2(sI-A)^{-1}B_1$.
- We would like to repeat the above by probing in channel two to pass information from channel two to channel one; so we will clearly need $\bar{n}$ to be at least as large as the relative degree of $C_1(sI-A)^{-1}B_2$.

So at this point, we define

$$\bar{n} := \max\{n_1 - 1, n_2 - 1, \text{r.d. of } C_2(sI-A)^{-1}B_1,\ \text{r.d. of } C_1(sI-A)^{-1}B_2\}. \tag{7.22}$$

Given the importance of the Markov parameters in the above analysis, we also define the vector

$$V_{i,j} := \begin{bmatrix} 0 \\ C_iB_j \\ \vdots \\ C_iA^{\bar{n}-1}B_j \end{bmatrix}, \quad i,j \in \{1,2\};$$

notice that both $V_{1,2}$ and $V_{2,1}$ are nonzero. We also define several matrices which come out of the above analysis, namely the two $(\bar{n}+1)\times(\bar{n}+1)$ matrices:

$$S := \begin{bmatrix} 1 & 0 & 0 & \cdots & 0 \\ 1 & 1 & 1 & \cdots & 1 \\ 1 & 2 & 2^2 & \cdots & 2^{\bar{n}} \\ & & \vdots & & \\ 1 & \bar{n} & \bar{n}^2 & \cdots & \bar{n}^{\bar{n}} \end{bmatrix}, \quad H(h) := \text{diag}\{1, h, \frac{h^2}{2!}, \ldots, \frac{h^{\bar{n}}}{\bar{n}!}\},$$

as well as two observability-like matrices and two vectors of sampled outputs:

$$\mathcal{O}_i := \begin{bmatrix} C_i \\ C_iA \\ \vdots \\ C_iA^{\bar{n}} \end{bmatrix} \in \mathbf{R}^{(\bar{n}+1)\times n}, \quad \mathcal{Y}_i(t) := \begin{bmatrix} y_i(t) \\ y_i(t+h) \\ \vdots \\ y_i(t+\bar{n}h) \end{bmatrix}, \quad i = 1, 2.$$

In the following, $\rho \in (0,1)$ is a scaling factor used when carrying out probing. Furthermore, we let $e_i \in \mathbf{R}^2$ denote the ith orthonormal vector.

Lemma 1 ([3]) *For every $\rho \in (0,1)$ and $\tilde{h} \in (0,1)$ there exists a constant $\tilde{\gamma} > 0$ so that for every $t_0 \in \mathbf{R}$, $x_0 \in \mathbf{R}^n$, $h \in (0, \tilde{h})$, $\bar{u} \in \mathbf{R}^2$ and $\delta \in \mathbf{R}$:*
(i) The solution of (7.1) with

$$u(t) = \begin{cases} (1+\rho)\bar{u} & t \in [t_0, t_0 + \bar{n}h) \\ (1-\rho)\bar{u} & t \in [t_0 + \bar{n}h, t_0 + 2\bar{n}h) \end{cases}$$

satisfies, for $i = 1, 2$:

$$\|H(h)^{-1}S^{-1}[\frac{\rho - 1}{2\rho}\mathcal{Y}_i(t_0) + \frac{\rho + 1}{2\rho}\mathcal{Y}_i(t_0 + \bar{n}h)] - \mathcal{O}_i x_0\| \le \tilde{\gamma} h(\|x_0\| + \|\bar{u}\|).$$

(ii) The solution of (7.1) with

$$u(t) = \begin{cases} \bar{u} + e_j\delta & t \in [t_0, t_0 + \bar{n}h) \\ \bar{u} - e_j\delta & t \in [t_0 + \bar{n}h, t_0 + 2\bar{n}h) \end{cases}$$

satisfies, for $i, j \in \{1, 2\}$:

$$\|H(h)^{-1}S^{-1}[\mathcal{Y}_i(t_0) - \mathcal{Y}_i(t_0 + \bar{n}h)] - 2V_{i,j}\delta\| \le \tilde{\gamma} h(\|x_0\| + \|\bar{u}\| + \|\delta\|).$$

To see how this result can be used, suppose that we have an initial estimate of the optimal control signal u^{opt} at time kT, which we label $\hat{u}^{opt}(kT)$. In the **first step**, the goal is to estimate $\xi^i(kT) = \begin{bmatrix} I_{n_i} & 0 \end{bmatrix} \mathcal{O}_i x(kT)$ in channel i. To proceed, we set

$$u(t) = \begin{cases} (1+\rho)\hat{u}^{opt}(kT) & t \in [kT, kT + \bar{n}h) \\ (1-\rho)\hat{u}^{opt}(kT) & t \in [kT + \bar{n}h, kT + 2\bar{n}h). \end{cases}$$

Then part (i) of Lemma 1 says that we can obtain a good estimate of $\xi^i(kT)$ in channel i very easily via

$$\hat{\xi}^i(kT) := \begin{bmatrix} I_{n_i} & 0 \end{bmatrix} H(h)^{-1}S^{-1}[\frac{\rho - 1}{2\rho}\mathcal{Y}_i(kT) + \frac{\rho + 1}{2\rho}\mathcal{Y}_i(kT + \bar{n}h)].$$

Of course, $\hat{\xi}^i(kT)$ is available only in channel i.

In the second step, the goal is pass the estimate of $\xi^1(kT)$ from channel one to channel two and the estimate of $\xi^2(kT)$ from channel two to channel one. Assuming that $n_1 > 0$, we can probe in the first channel with the first element of $\hat{\xi}^1(kT)$, which we label $[\hat{\xi}^1(kT)]_1$:

$$u(t) = \hat{u}^{opt}(kT) + \begin{cases} \rho e_1[\hat{\xi}^1(kT)]_1 & t \in [kT + 2\bar{n}h, kT + 3\bar{n}h) \\ -\rho e_1[\hat{\xi}^1(kT)]_1 & t \in [kT + 3\bar{n}h, kT + 4\bar{n}h); \end{cases}$$

then the probing takes place only in the first channel using information available in that channel. We can then form an estimate of $[\xi^1(kT)]_1$ in channel two, which we label $[\hat{\hat{\xi}}^1(kT)]_1$, using Lemma 1 (ii) and the fact that $V_{2,1} \neq 0$:

$$\frac{1}{2\rho} H(h)^{-1} S^{-1} [\mathcal{Y}_2(kT + 2\bar{n}h) - \mathcal{Y}_2(kT + 3\bar{n}h)] \approx V_{2,1}[\xi^1(kT)]_1,$$

which means that we should define the first element of $\hat{\hat{\xi}}^1(kT)$ by

$$\begin{aligned} [\hat{\hat{\xi}}^1(kT)]_1 &:= \underbrace{\frac{1}{2\rho}(V_{2,1}{}^T V_{2,1})^{-1} V_{2,1}{}^T H(h)^{-1} S^{-1}}_{=:\tilde{V}_{2,1}} \times [\mathcal{Y}_2(kT + 2\bar{n}h) - \mathcal{Y}_2(kT + 3\bar{n}h)] \\ &\approx [\hat{\xi}^1(kT)]_1. \end{aligned}$$

Of course, if $n_1 > 1$ we can repeat this experiments to obtain an estimate of the second element of $\xi^1(kT)$ in channel two as well: we set

$$u(t) = \hat{u}^{opt}(kT) + \begin{cases} \rho e_1[\hat{\xi}^1(kT)]_2 & t \in [kT + 4\bar{n}h, kT + 5\bar{n}h) \\ -\rho e_1[\hat{\xi}^1(kT)]_2 & t \in [kT + 5\bar{n}h, kT + 6\bar{n}h), \end{cases}$$

and then obtain an estimate of $[\xi^1(kT)]_2$ in channel two:

$$[\hat{\hat{\xi}}^1(kT)]_2 := \tilde{V}_{2,1} \times [\mathcal{Y}_2(kT + 4\bar{n}h) - \mathcal{Y}_2(kT + 5\bar{n}h)] \approx [\hat{\xi}^1(kT)]_2.$$

We can carry out the same procedure for the remaining elements of $\hat{\xi}^1(kT)$, thereby yielding an estimate $\hat{\hat{\xi}}^1(kT)$ of $\xi^1(kT)$ in channel two; after that we can probe in channel two to pass the estimate of $\xi^2(kT)$ from channel two to channel one, thereby yielding an estimate of it in channel one, which we label $\hat{\hat{\xi}}^2(kT)$. In channel i, we now have sufficient information to obtain an estimate of $x(kT)$: in channel one, this is given by

$$\mathcal{O}_{est}^{-1} \begin{bmatrix} \hat{\xi}^1(kT) \\ \hat{\hat{\xi}}^2(kT) \end{bmatrix},$$

while in channel two, this is given by

$$\mathcal{O}_{est}^{-1} \begin{bmatrix} \hat{\hat{\xi}}^1(kT) \\ \hat{\xi}^2(kT) \end{bmatrix}.$$

This allows us to form a good estimate of $F^i x(kT)$ for use during the next period $[(k+1)T, (k+2)T)$.

At this point, we are ready to provide a precise definition of the controller. Before proceeding, we define $\tilde{V}_{1,2}$ in a manner analogous to that of $\tilde{V}_{2,1}$ defined above.

7.4.3 The Final Controller and the Main Result

With $\bar{n}$ defined in (7.22), fix $h > 0$ and $\rho > 0$. Since we saw above that $\bar{n}h$ plays an important role, we give it a name:

$$\bar{h} := \bar{n}h.$$

It turns out that we need

- two intervals of length $\bar{h}$ to estimate $\hat{\xi}^i(kT)$ in channel i;
- $2n$ intervals of length $\bar{h}$ to pass this information to the other channel;
- one interval of length $\bar{h}$ to compute the updated control law.[4]

We subsequently set $\ell := (2n+3)\bar{n}$ and set the controller period to be $T := \ell h$.

With $\hat{u}^{opt}(0) = 0$, we define the controller via three parts— for $k \in \mathbf{Z}^+$:

Step 1: Construct $\hat{\xi}^i(kT)$ in channel i on $[kT, kT + 2\bar{h})$.

Set

$$u(t) = \begin{cases} (1+\rho)\hat{u}^{opt}(kT) & t \in [kT, kT + \bar{h}) \\ (1-\rho)\hat{u}^{opt}(kT) & t \in [kT + \bar{h}, kT + 2\bar{h}), \end{cases} \tag{7.23}$$

and define

$$\hat{\xi}^i(kT) := \begin{bmatrix} I_{n_i} & 0 \end{bmatrix} H(h)^{-1} S^{-1} [\frac{\rho - 1}{2\rho} \mathcal{Y}_i(kT) + \frac{\rho + 1}{2\rho} \mathcal{Y}_i(kT + \bar{h})], \quad i = 1, 2. \tag{7.24}$$

Step 2: Estimate $\hat{\hat{\xi}}^i(kT)$ on $[kT + 2\bar{h}, kT + (2n+2)\bar{h})$.

We start by probing in channel one using the n_1 elements of $\hat{\xi}^1(kT)$ in sequence:

$$u(t) = \hat{u}^{opt}(kT) + \begin{cases} \rho e_1 [\hat{\xi}^1]_1 (kT) & t \in [kT + 2\bar{h}, kT + 3\bar{h}) \\ -\rho e_1 [\hat{\xi}^1]_1 (kT) & t \in [kT + 3\bar{h}, kT + 4\bar{h}) \\ \vdots & \vdots \\ \rho e_1 [\hat{\xi}^1]_{n_1} (kT) & t \in [kT + 2n_1\bar{h}, kT + (2n_1+1)\bar{h}) \\ -\rho e_1 [\hat{\xi}^1]_{n_1} (kT) & t \in [kT + (2n_1+1)\bar{h}, kT + (2n_1+2)\bar{h}) \end{cases} \tag{7.25}$$

[4] This time interval can be made shorter, but it would complicate the notation.

and then form an estimate $\hat{\hat{\xi}}^1(kT)$ of $\xi^1(kT)$ in the second channel:

$$\hat{\hat{\xi}}^1(kT) := \text{diag}\,\underbrace{\{\tilde{V}_{2,1},\ldots,\tilde{V}_{2,1}\}}_{n_1 \text{ copies}} \times \begin{bmatrix} \mathcal{Y}_2(kT+2\bar{h}) - \mathcal{Y}_2(kT+3\bar{h}) \\ \vdots \\ \mathcal{Y}_2(kT+2n_1\bar{h}) - \mathcal{Y}_2(kT+(2n_1+1)\bar{h}) \end{bmatrix}. \tag{7.26}$$

Next, we probe in channel two using the n_2 elements of $\hat{\xi}^2(kT)$ in sequence:

$$u(t) = \hat{u}^{opt}(kT) + \begin{cases} \rho e_2[\hat{\xi}^2]_1(kT) & t \in [kT+(2n_1+2)\bar{h}, kT+(2n_1+3)\bar{h}) \\ -\rho e_2[\hat{\xi}^2]_1(kT) & t \in [kT+(2n_1+3)\bar{h}, kT+(2n_1+4)\bar{h}) \\ \vdots & \vdots \\ \rho e_2[\hat{\xi}^2]_{n_2}(kT) & t \in [kT+2n\bar{h}, kT+(2n+1)\bar{h}) \\ -\rho e_2[\hat{\xi}^2]_{n_2}(kT) & t \in [kT+(2n+1)\bar{h}, kT+(2n+2)\bar{h}) \end{cases} \tag{7.27}$$

and then form an estimate $\hat{\hat{\xi}}^2(kT)$ of $\xi^2(kT)$ in the first channel:

$$\hat{\hat{\xi}}^2(kT) := \text{diag}\,\underbrace{\{\tilde{V}_{1,2},\ldots,\tilde{V}_{1,2}\}}_{n_2 \text{ copies}} \times \begin{bmatrix} \mathcal{Y}_1(kT+(2n_1+2)\bar{h}) - \mathcal{Y}_1(kT+(2n_1+3)\bar{h}) \\ \vdots \\ \mathcal{Y}_1(kT+2n\bar{h}) - \mathcal{Y}_1(kT+(2n+1)\bar{h}) \end{bmatrix}. \tag{7.28}$$

We now can form an estimate of $x(kT)$ in each channel:

$$\hat{x}^1(kT) := \mathcal{O}_{est}^{-1} \begin{bmatrix} \hat{\xi}^1(kT) \\ \hat{\hat{\xi}}^2(kT) \end{bmatrix}, \tag{7.29}$$

$$\hat{x}^2(kT) := \mathcal{O}_{est}^{-1} \begin{bmatrix} \hat{\hat{\xi}}^1(kT) \\ \hat{\xi}^2(kT) \end{bmatrix}. \tag{7.30}$$

Step 3: Form the updated control law on $[kT+(2n+2)\bar{h}, kT+(2n+3)\bar{h})$.
Set

$$u(t) = \hat{u}^{opt}(kT), \quad t \in [kT+(2n+2)\bar{h}, kT+(2n+3)\bar{h}), \tag{7.31}$$

and define the updated control signal for use during the following period:

$$\hat{u}^{opt}((k+1)T) = \begin{bmatrix} F_1\hat{x}^1(kT) \\ F_2\hat{x}^2(kT) \end{bmatrix}. \tag{7.32}$$

The behaviour of the above controller (7.23)–(7.32) is governed by its state

$$\begin{bmatrix} \hat{x}^1(kT) \\ \hat{x}^2(kT) \end{bmatrix},$$

which is initialized to zero to reflect the choice of $\hat{u}^{opt}(0) = 0$. To this end, we define the sampled-data closed-loop state by

$$x_{sd}(t) := \begin{bmatrix} x(t) \\ \hat{x}^1(kT) \\ \hat{x}^2(kT) \end{bmatrix}, \quad t \in [kT, (k+1)T).$$

For a given initial condition x_0 and with $t_0 = 0$, we let $J(x_0)$ denote the corresponding cost

$$J(x_0) = \int_0^{\infty} [x(t)^T Qx(t) + u(t)^T Ru(t)]dt.$$

Theorem 2 *For every $\varepsilon > 0$, there exists a period $T > 0$, a scaling factor $\rho > 0$ and constants $\gamma > 0$ and $\lambda < 0$, so that when the corresponding controller (7.23)–(7.32) is applied to the plant (7.1) with $t_0 = 0$:*

(i) The closed-loop system is exponentially stable in the sense that, for every $x_0 \in \mathbf{R}^n$, we have

$$\|x_{sd}(t)\| \le \gamma e^{\lambda t}\|x_0\|, \quad t \ge 0;$$

(ii) It provides near-optimal performance in the sense that, for every $x_0 \in \mathbf{R}^n$, we have

$$J(x_0) \le J^{opt}(x_0) + \varepsilon\|x_0\|^2.$$

Proof This result is a special case of the Theorem 1 of [3].

□

Remark 1 Given an $\varepsilon > 0$, we can achieve our objective by carrying out the above controller design and making $\rho > 0$, $h > 0$ and $\frac{h}{\rho}$ sufficiently small.

Remark 2 Under the assumptions in this section, the decentralized plant may have unstable DFMs, but it has no QFMs. In Chap. 5, we concluded that in this situation we can always stabilize the plant using a linear periodic controller; here, we have proven that we can obtain near-optimal LQR performance as well!

Remark 3 Since the controller is linear (though periodic), it is easy to see that the closed-loop system is noise tolerant, in the sense that if we inject noise at both plant controller interfaces, then the map from the noise to the signals $x(t)$ and $u(t)$ has a finite gain in the induced ∞-norm sense. For the same reason, we can use the Small-Gain Theorem to verify that the closed-loop system maintains stability in the presence of a small amount of unmodelled dynamics.

7.4.4 An Example

Consider the following system

$$\dot{x} = \begin{bmatrix} -1 & 0 & 0 \\ 0 & 1 & 0 \\ 0 & 0 & 2 \end{bmatrix} x + \begin{bmatrix} 1 \\ 1 \\ 0 \end{bmatrix} u_1 + \begin{bmatrix} 0 \\ 0 \\ 1 \end{bmatrix} u_2$$
$$y_1 = \begin{bmatrix} 0 & 0 & 1 \end{bmatrix} x$$
$$y_2 = \begin{bmatrix} 3 & -1 & 0 \end{bmatrix} x.$$

It is easy to confirm that the system is centrally controllable and centrally observable; furthermore, the corresponding transfer function is

$$G(s) = \begin{bmatrix} 0 & \frac{1}{s-2} \\ \frac{2(s-2)}{(s+1)(s-1)} & 0 \end{bmatrix},$$

which means that the associated graph is strongly connected. Since this plant clearly has a DFM at $s = 2$ (see Sect. 5.2), there is no LTI controller which provides closed-loop stability. However, by Theorem 2 there is a linear periodic sampled-data controller which provides near-optimal LQR performance. Here we adopt the performance index of

$$\int_0^\infty [x(t)^T x(t) + u(t)^T u(t)] dt;$$

it turns out that the optimal controller is state feedback:

$$u = \begin{bmatrix} 0.0000 & -2.7321 & 0 \\ 0 & 0 & -4.2361 \end{bmatrix} x,$$

with an optimal cost of

$$x_0^T \begin{bmatrix} 0.5000 & -0.5000 & 0 \\ -0.5000 & 3.2321 & 0 \\ 0 & 0 & 4.2361 \end{bmatrix} x_0.$$

So our objective is to design a stabilizing sampled-data controller which provides a level of performance which is close to this.

We proceed by constructing the controller which was proposed in this section. We first need to ascertain some key quantities:

(i) First we need to find n_1 and n_2. To this end, observe that the observability matrices associated with (C_1, A) and (C_2, A) are

$$\mathcal{O}(C_1, A) := \begin{bmatrix} C_1 \\ C_1 A \\ C_1 A^2 \end{bmatrix} = \begin{bmatrix} 0 & 0 & 1 \\ 0 & 0 & 2 \\ 0 & 0 & 4 \end{bmatrix}$$

and

$$\mathcal{O}(C_2, A) := \begin{bmatrix} C_2 \\ C_2 A \\ C_2 A^2 \end{bmatrix} = \begin{bmatrix} 3 & -1 & 0 \\ -3 & -1 & 0 \\ 3 & -1 & 0 \end{bmatrix},$$

respectively. We see that the first $n_1 = 1$ columns of $\mathcal{O}(C_1, A)$ and the first $n_2 = 2$ columns of $\mathcal{O}(C_2, A)$ are linearly independent, giving rise to

$$\xi^1(t) = C_1 x(t) = y_1(t),$$

$$\xi^2(t) = \begin{bmatrix} C_2 \\ C_2 A \end{bmatrix} x(t),$$

and

$$\mathcal{O}_{est} := \begin{bmatrix} \begin{bmatrix} I_{n_1} & 0 \end{bmatrix} & 0 \\ 0 & \begin{bmatrix} I_{n_2} & 0 \end{bmatrix} \end{bmatrix} \begin{bmatrix} \mathcal{O}(C_1, A) \\ \mathcal{O}(C_2, A) \end{bmatrix} = \begin{bmatrix} 0 & 0 & 1 \\ 3 & -1 & 0 \\ -3 & -1 & 0 \end{bmatrix},$$

which is invertible. Hence,

$$\mathcal{O}_{est}^{-1} \begin{bmatrix} \xi^1(t) \\ \xi^2(t) \end{bmatrix} = x(t).$$

(ii) Next we need to compute $\bar{n}$—using (7.22) we have

$$\begin{aligned} \bar{n} &:= \max\{n_1 - 1, n_2 - 1, \text{r.d. of } C_2(sI - A)^{-1}B_1, \text{ r.d. of } C_1(sI - A)^{-1}B_2\} \\ &= \max\{0, 1, 1, 1\} \\ &= 1. \end{aligned}$$

(iii) Now we need to compute $V_{1,2}$ and $V_{2,1}$:

$$V_{1,2} = \begin{bmatrix} 0 \\ C_1 B_2 \end{bmatrix} = \begin{bmatrix} 0 \\ 1 \end{bmatrix}, \quad V_{2,1} = \begin{bmatrix} 0 \\ C_2 B_1 \end{bmatrix} = \begin{bmatrix} 0 \\ 2 \end{bmatrix}.$$

Now we construct the controller outlined in this section. We have $\bar{h} = \bar{n}h = h$, $\ell = (2n + 3)\bar{n} = 9$ and $T = \ell h = 9h$. From the description of the controller, on each period $[kT, (k + 1)T)$ we do the following:

- In Step 1, which lasts $2\bar{h} = 2h$ units of time, carry out estimation of $\xi^1(kT)$ in channel one and an estimation of $\xi^2(kT)$ in channel two.

- In Step 2, which lasts $2n\bar{h} = 6h$ units of time, we probe in channel one for $2h$ units of time to pass an estimate of $\xi^1(kT)$ to channel two, and then we probe in channel two for $4h$ units of time to pass an estimate of $\xi^2(kT)$ to channel one.
- In Step 3, which lasts $\bar{h} = h$ units of time, we form the updated control signal $\hat{u}^{opt}((k+1)T)$.

For a given $\rho > 0$, for small h (and therefore T) stability is attained; to get performance close to the optimal we need both ρ and T to go to zero in just the right way.

We have carried out a simulation for the case of $\rho = 0.1$, $h = 0.002$ and $T = 0.018$. With $x(0) = \begin{bmatrix} 1 & 1 & 1 \end{bmatrix}^T$ and $\hat{u}^{opt}(0) = 0$, we have carried out a simulation and displayed the results in Figs. 7.1 and 7.2; in each plot, we have compared them to the optimal response. We see that the response of x is nearly optimal, while the control signal looks like the optimal control signal with some dither added. In Fig. 7.3, we have provided a closeup of the control signal during the first ten periods—it naturally remains close to $\hat{u}^{opt}(0)$ during the first period but then tracks the optimal control signal after that. We can make that difference smaller by reducing the size of ρ and h.

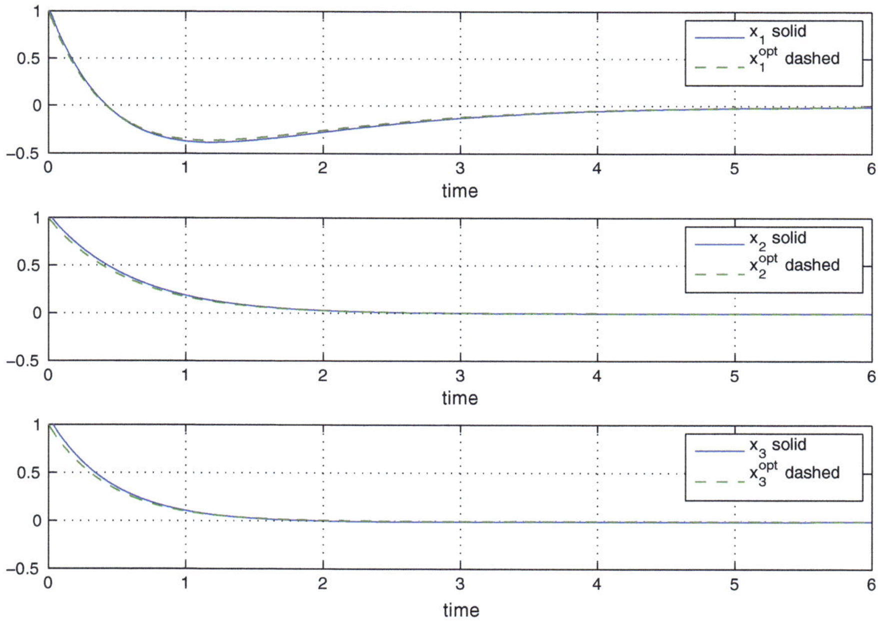

Fig. 7.1 The closed-loop state response

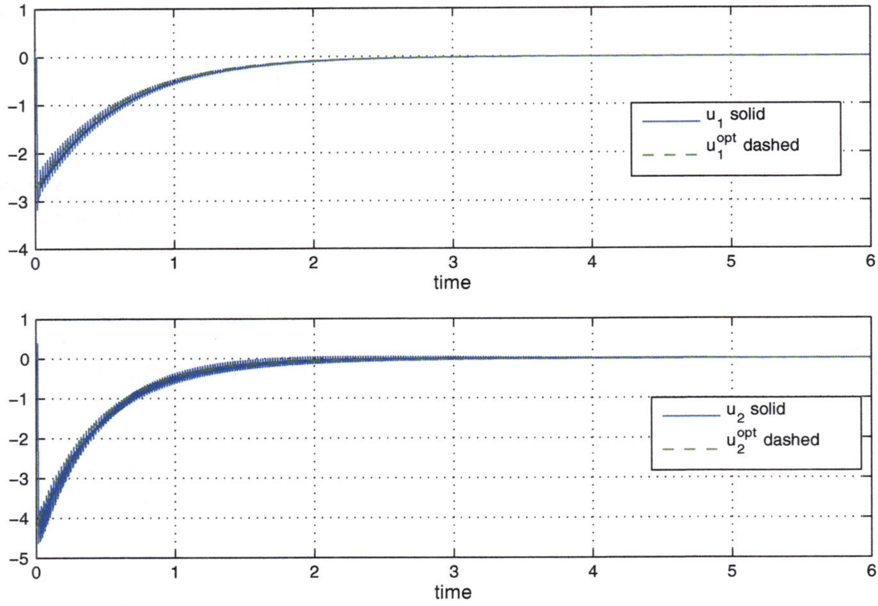

Fig. 7.2 The closed-loop control signal response

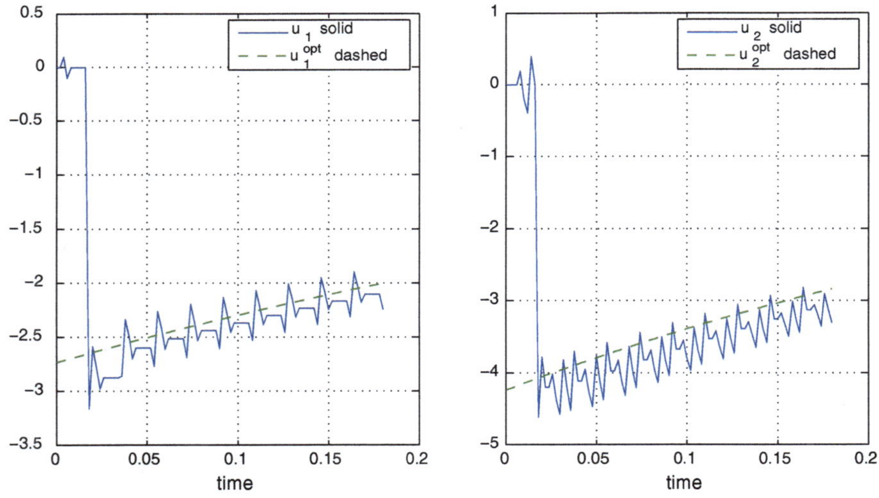

Fig. 7.3 The closeup of the control signal

7.5 The General Strongly Connected Case

In the previous section we made two simplifying assumptions: there are two channels, i.e. $\nu = 2$, and the channels are all scalar-valued. Here, we briefly explain how to remove these assumptions.[5]

7.5.1 *The Case of $\nu > 2$*

So suppose that $\nu > 2$. One of the key aspects of the approach is that information is passed from each channel to the others. In the case of $\nu = 2$, this is possible since strong connectedness ensures that the transfer functions

$$C_1(sI - A)^{-1}B_2 \text{ and } C_2(sI - A)^{-1}B_1$$

are both nonzero. In the case of $\nu > 2$, this generalizes into the requirement that

$$C_i(sI - A)^{-1}B_j \neq 0 \text{ for every } i \neq j; \tag{7.33}$$

however, strong connectedness does not ensure that this is the case. Fortunately, it turns out that if this is not the case, then the system can be easily regularized. To proceed, we first define the set of admissible decentralized feedback gains by

$$\mathcal{K} :=:= \{K \in \mathbf{R}^{m\times r} : K = \text{diag}\{K_1, \ldots, K_\nu\} \text{ with } K_i \in \mathbf{R}^{m_i\times r_i}\}.$$

The regularization is carried out by applying some output feedback to the system. Before stating this, we need a definition.

Definition 1 A **property** π is a function $\pi : \mathcal{K} \to \{0, 1\}$, where $\pi(K) = 1$ (or 0) means that π holds (or fails) at K. A property π holds for **almost all** $K \in \mathcal{K}$ if there exists a **nonzero** polynomial ϕ for which π holds at x if and only if $\phi(x) \neq 0$.

Proposition 2 (Propositions 1 and 2 of [1]) *For almost all $K \in \mathcal{K}$, for every $i, j, \in \{1, \ldots, \nu\}$ the transfer function $C_i(sI - A - BKC)^{-1}B_j$ is not identically zero.*

Hence, if the graph associated with the original plant (7.1) does not satisfy (7.33), then we simply choose a regularizing $K \in \mathcal{K}$ which Proposition 2 asserts to exist, and set

---

[5] This section is derived, in part, from [3].

$$u = Ky + u^n;$$

if we apply this control law to the plant (7.1), then we end up with a slightly modified system

$$\begin{aligned} \dot{x} &= \underbrace{(A + BKC)}_{=:A_n} x + Bu^n \\ y &= Cx \end{aligned}$$

which has a complete graph. Of course, the performance index and the optimal control law has to be adjusted slightly as well; see [3] for details.

7.5.2 Non-scalar Channels

In the previous section, it is assumed that all channels are scalar-valued; this is especially important for the passing of information from one channel to another. It turns out that it is easy to convert a system with non-scalar channels to one with scalar channels. First assume that the regularization carried out in the previous sub-section has already been carried out so that the corresponding graph is complete. Hence, for every $i, j \in \{1, \dots, \nu\}$, the transfer function $C_i(sI - A)^{-1}B_j$ is not identically zero; it is not hard to prove that for almost all $v_i \in \mathbf{R}^{r_i}$ and $\bar{v}_j \in \mathbf{R}^{m_i}$, we have that the scalar transfer function $v_i^T C_i(sI - A)^{-1}B_j\bar{v}_j$ is not identically zero either. Hence, during the probing process of passing information from one channel to another, we use a new plant output

$$\bar{y}_i := \underbrace{v_i^T C_i}_{=:\bar{C}_i} x;$$

if we let $\bar{u}_i$ be a new scalar input, set the actual input to

$$u_i = \bar{v}_i \bar{u}_i$$

and define $\bar{B}_i := B_i \bar{v}_i$ accordingly, then the plant equation becomes

$$\begin{aligned} \dot{x} &= Ax + \sum_{i=1}^{\nu} \bar{B}_i \bar{u}_i \\ \bar{y}_i &= \bar{C}_i x, \ i = 1, \dots, \nu, \end{aligned}$$

for which all channels are scalar. Hence, we are effectively back in the scalar channel case. The low-level details of how to make this work are given in [3].

7.6 Summary

The fact that there is a gap between the optimal performance in the decentralized setting versus the centralized setting has been known for some time, certainly as far back as Witsenhausen [6], which was in the stochastic setting. The LQR example provided in Sect. 7.2 is a much simpler version of the example presented in Miller [2] and demonstrates that the gap can be arbitrarily large.

Finding the optimal controller in the decentralized setting is known to be quite difficult. Roughly speaking, there are two general lines of inquiry:

- In the first classical line of inquiry, the goal is to ascertain classes of systems where either the optimal controller is LTI or the objective is to find the optimal LTI controller; it turns out that a lot of plant structure is required for this to work. This is not the focus here; a good reference for the approach is that of Rotkowitz and Lall [4] as well as the introduction of Miller and Davison [3].
- The second line of inquiry is to ascertain classes of systems where the optimal controller is possibly non-LTI, but for which it is possible to compute a (near) optimal controller; indeed, it is especially interesting if there does not even exist a stabilizing LTI controller, let alone one which provides good performance. This is the focus of the last two sections of this chapter and is centred on work done by two of the book's co-authors Miller and Davison in the LQR setting [3]; the key assumption is the strong connectedness of the system graph. This has been extended to the H_∞ setting in Rangananthan and Miller [5]; the problem is more difficult there since the goal is to minimize the effect of an exogenous input rather than to minimize the transient effect of the initial condition.

7.7 Problems

1. It turns out that the bound provided by Proposition 1 is not very tight. Prove that

$$\limsup_{k\to\infty} J_k^{opt}(\begin{bmatrix} 1 \\ 0 \end{bmatrix}) < \infty.$$

Hint: While a simulation indicates that $\limsup_{k\to\infty} J_k^{opt}(\begin{bmatrix} 1 \\ 0 \end{bmatrix}) = 1$, we will be able to prove a slightly weaker bound. Based on a numerical analysis, a crude approximation to the optimal feedback matrix is

$$\hat{F}(k) := \begin{bmatrix} -2 & -(k-2)^{1/2} \\ -(k-2)^{1/2} & -k+1 \end{bmatrix},$$

restricted to $k > 2$. First show that this places the closed-loop poles, i.e. the eigenvalues of $A(k) + B(k)\hat{F}(k)$, exactly at $\{-1, -k\}$. Second of all, with an initial condition of $x(0) = \begin{bmatrix} 1 \\ 0 \end{bmatrix}$ and $u(t) = \hat{F}(k)x(t)$, obtain an upper bound on quadratic cost as a function of k, and then take a limit as $k \to \infty$.

2. For the LQR example provided in Sect. 7.4.4: with the choice of h given there, simplify the description of the controller (7.23)–(7.32) as far as possible.

3. Consider the system with a transfer function from u to y of

$$G(s) = \begin{bmatrix} 0 & 0 & g_1(s) \\ g_2(s) & 0 & 0 \\ 0 & g_3(s) & 0 \end{bmatrix}$$

with $g_1(s)$, $g_2(s)$ and $g_3(s)$ all nonzero, strictly proper, transfer functions. Now consider the regularizing output feedback of

$$u = \underbrace{\begin{bmatrix} k_1 & 0 & 0 \\ 0 & k_2 & 0 \\ 0 & 0 & k_3 \end{bmatrix}}_{=:K} y + u^n$$

with $k_i \in \mathbf{R}$. Show that the graph associated with the closed-loop transfer function from u^n to y is complete if and only if $k_1 k_2 k_3 \neq 0$ (i.e. all of the $k_i's$ are nonzero).

Hint: Examine the adjoint matrix of $I - GK$.

4. For the example in the previous question, prove that if $k_1 k_2 k_3 \neq 0$, then the relative degree of the transfer function from u_0 to y is the sum of the relative degrees of $g_1(s)$, $g_2(s)$ and $g_3(s)$. Hence, the relative degree is affected by the regularization process; this turns out to be important in the H_∞ setup of [5].

References

1. L. Kobayashi, T. Yoshikawa, Graph-theoretic approach to controllability and localizability of decentralized control systems. IEEE Trans. Autom. Control **27**, 1096–1108 (1982)
2. D.E. Miller, A comparison of LQR optimal performance in the decentralized and centralized settings. IEEE Trans. Autom. Control **61**, 2308–2311 (2016)
3. D.E. Miller, E.J. Davison, Near optimal LQR control in the decentralized setting. IEEE Trans. Autom. Control **59**, 327–340 (2014)
4. M. Rotkowitz, S. Lall, A characterization of convex problems in decentralized control. IEEE Trans. Autom. Control **51**, 274–286 (2006)

5. T. Ranganathan, D.E. Miller, Near optimal H_∞ performance in the decentralized setting. Math. Control Sign. Syst. **30**(2) (2018)
6. H.S. Witsenhausen, A counterexample in stochastic optimum control. SIAM J. Control **6**(1), 131–147 (1968)

Index

E. J. Davison et al., *Decentralized Control of Large-Scale Systems*, https://doi.org/10.1007/978-1-4419-6014-6